W. Nachtigall

Biological Mechanisms of Attachment

The Comparative Morphology and Bioengineering of Organs for Linkage, Suction, and Adhesion

Translated by M. A. Biederman-Thorson

With 63 Plates comprising 721 Illustrations

Springer-Verlag Berlin Heidelberg GmbH 1974

Professor Dr. Werner Nachtigall
Zoologisches Institut der Universität des Saarlandes
D-6600 Saarbrücken, W. Germany

Marguerite A. Biederman-Thorson, Ph. D.
The Old Marlborough Arms, Combe
Oxford, England

ISBN 978-3-642-85777-5 ISBN 978-3-642-85775-1 (eBook)
DOI 10.1007/978-3-642-85775-1

Originally published by Springer-Verlag Berlin Heidelberg New York in 1974
Softcover reprint of the hardcover 1st edition 1974

Library of Congress Catalog Card Number
73-17936.
Typesetting: Zechnersche Buchdruckerei, Speyer.

Preface

Bioengineering is the branch of biology which applies the methods of engineering and physics to the study of biological phenomena, and the vocabulary of technology to describe them. Particularly with respect to the mechanics of movement and other physiological processes, the advantages of this approach are obvious. But other fields of study also reveal new insights when biotechnical research methods are applied, and one of these is the comparative morphology of biological structures. At the very least, description in technical terms permits complete, concise organization of a field of research, provides a means of describing biological forms in terms consistent with their function, and aids in working out interpretations based on structural design and functional anatomy.

It is from this point of view that the present book describes and discusses, on a comparative basis, biological mechanisms of attachment. Although these are among the simplest biological mechanisms, they are fascinating in their diversity. This presentation is, in a way, an outgrowth of that encyclopedic drive which is within every scientist. Over the years, whole libraries of books have been scanned as a result of this general curiosity; the library of the Munich Zoological Institute has proved a particularly valuable source of information. This little book is a much abbreviated distillation of the several thousand file cards which resulted from this urge to collect. Despite the brevity of presentation, its preparation—particularly the time-consuming organization of the plates—has extended over years, in such moments as could be spared from other work.

At a time when research and teaching concentrate more and more on particular topics and examples, a collection such as this may seem something of an anachronism. But it appears to me necessary that the biologist recall from time to time how extraordinarily varied are the possibilities which nature offers for solving any set of problems. This compendium of facts is intended to contribute to that end.

The text has been kept very concise; every important detail is elucidated in an illustration. Although only a selection of fundamental and typical structures is given, the collection amounts to no fewer than 721 figures. It was quite impossible to redraw the illustrations for this specialized work. Almost all of the pictures are reproductions from the literature, which I have made over the years. I thank the authors of these figures and the publishers of the books and journals in which they appeared for permission to use the reproductions in this collection. The sources of the figures are summarized in Section 4. To find the original sources of the corresponding descriptions in the text, compare the authors' names in Section 4 with the References.

The figures are cited in the text in parentheses. Explanations of the labels in the drawings, if not in the text, are to be found in the legends.

I thank Dr. Biederman-Thorson for her critical and sensitive translation of the German text, and Springer-Verlag for the most attractive presentation of this book.

Winter 1973 W. Nachtigall

Contents

Introduction

Of all the ways the connections of structural elements in engineering and biology might be classified, our purposes will be well served by the two dichotomies "releasable/permanent" and "rigid/flexible". Taken in all combinations, these specify four different kinds of coupling, illustrated below with common technological applications:

	Releasable	Permanent
Rigid	electrical plug and socket	riveted plates of a ship's hull
Flexible	tailor's hook and eye	the hinge of a door

These four fundamentally distinct categories provide a framework for discussion of analogous connecting mechanisms in animals and plants. In terms of the biological counterparts, the hinge category is of course represented by the flexible joints of the appendages of animals; these can take a great many different forms, and since much has been written about them elsewhere they are not treated in this book. Moreover, the two "releasable" categories are difficult to distinguish from one another where biological structures are concerned, since the elasticity of the materials involved makes the notions "rigid" and "flexible" quite relative. Therefore the many kinds of releasable connections are grouped here under a single main heading.

1. Rigid, Permanent Attachments

This category comprises many techniques: soft soldering, hard soldering, welding, gluing, mortaring, grouting, puttying, sewing, folding, splicing, tying, weaving, and attaching by means of screws, rivets, nails, tacks or pegs. All such methods can be grouped loosely according to function under four headings. Connections may be made

by means of special devices
by means of amorphous bonding material
by softening and re-hardening of material
by anchoring, interlacing, and interlocking.

Special connecting devices include screws, rivets, pins, and thread. Examples from the second group are soldering, gluing, and mortaring. In the third category, additional material, identical or similar to that of the parts to be joined, can be used; such methods include metal welding, thermal welding of plastics, and bonding of plastics by solvents. The last group, which consists of anchoring, rooting, tying, folding, and interlocking joints like those used in carpentry, is special in that neither a softening of the material nor any sort of additional connecting element need be used.

1.1 Special Connecting Devices

In nature this type of attachment does not occur. Never are two structural parts held together by a third, entirely different one corresponding to the nail or the screw. On the other hand, attachment by means of amorphous bonding material occurs quite commonly.

1.2 Amorphous Bonding Material

In general, the description of biological attachments as "rigid-permanent" should be viewed as somewhat less absolute than the same term in engineering. Reinforced concrete is made by pouring the concrete around an iron framework so as to embed it immovably and non-removably, but such a bond is "immobile" only under normal use and from the ordinary macroscopic point of view. Microscopically small displacements can occur, and an earthquake can destroy the connection and free the steel. The above term, then, applies only in the ordinary

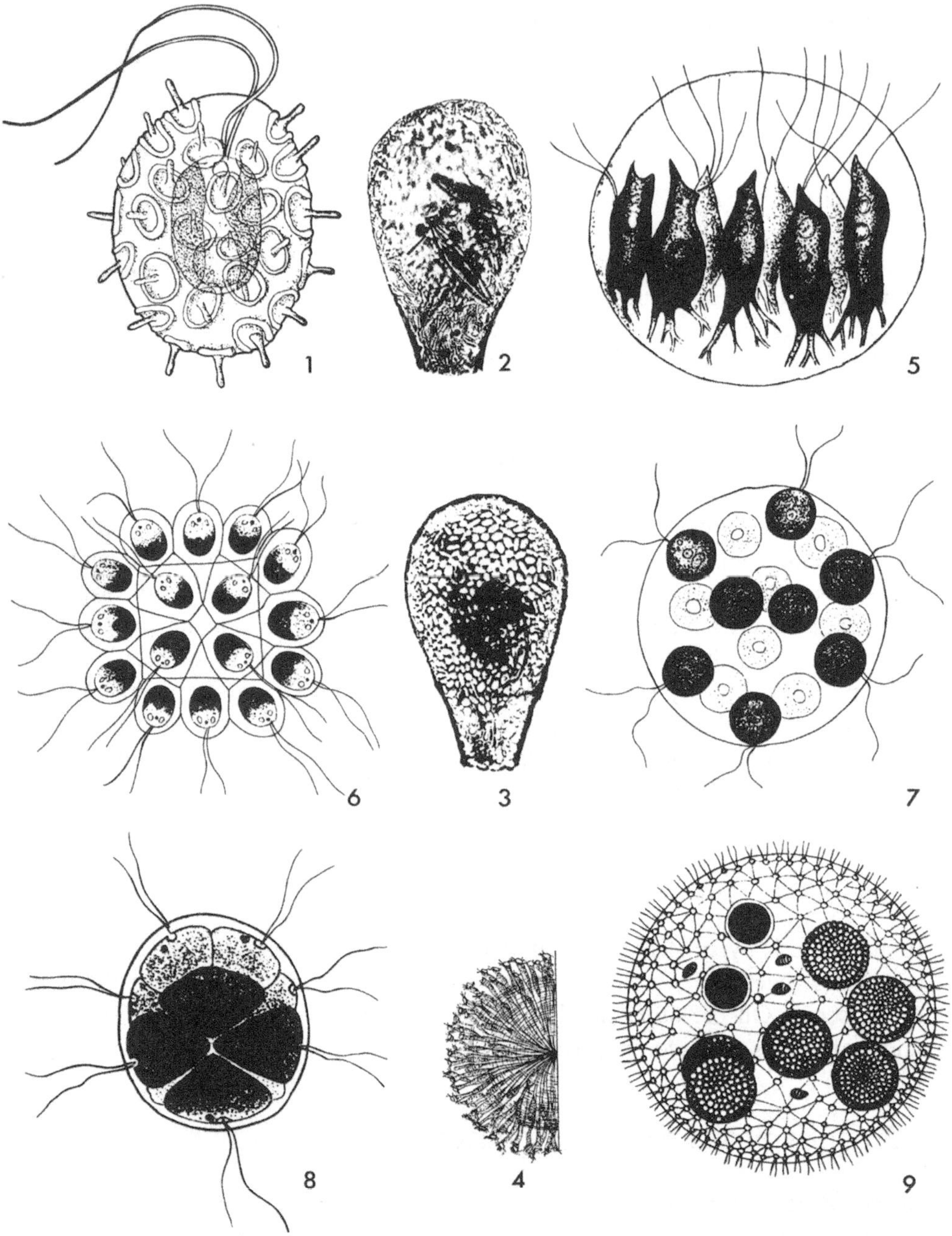

Plate 1 (Figs. 1–9)

1. A coccolithophorid
2. Difflugia
3. Nebela
4. Ophrydium versatile
5. Stephanosphaera pluvialis
6. Gonium pectorale
7. Eudorina elegans
8. Pandorina morum
9. Volvox aureus

situation. In nature, as well, if a connection is deformed or destroyed by strong forces, normal working conditions no longer prevail. For example, the connection between the calcareous plates and the amorphous matrix in the membranes surrounding marine flagellates of the family Coccolithophoridae (1) comes under the heading "rigid-permanent." It is true that one can deform this membranous case by the pressure of a cover slip, tear it apart with dissection needles, or dissolve out the little plates with acid; but under the conditions of its ordinary function the connection is firm and is never broken. The cases of the coccolithophorids are also a good example of connection by means of an amorphous substance. In this respect the calcareous plates in the membrane correspond closely to the tiles in a Dutch-tile stove, which are laid in a bed of mortar. The functions, though, are different: protection in nature, heat storage in technology.

These calcareous platelets are secreted by the animal itself. Other microorganisms, on the other hand, provide themselves with mechanical protection by embedding hard foreign bodies (grains of sand) in the organic (hyaline) cases which surround them; examples of these are the amoebas Difflugia (2) and Nebela (3). Similarly, caddis-fly larvae such as those of the genera Limnophilus and Phryganea stick pebbles, snail shells, and bits of leaves and twigs onto their sheaths of organic fabric, where they become firmly bound. A shapeless, gelatinous matrix joins individual peritrich ciliates Ophrydium versatile (4) into a colony; the long conical cells are arranged radially in a sphere of gelatine. Lower algae, phytomonad flagellates which bear chromatophores, have a number of genera in which the single cells are embedded in a gelatinous mass shaped like a sheet or a solid or hollow sphere; these include Stephanosphaera (5), Gonium (6), Eudorina (7), Pandorina (8), and Volvox (9).

In hyaline cartilage, the cells are firmly and inextricably embedded in a more or less homogeneous cartilaginous matrix. The leaves of horse-chestnut buds are stuck together by a resinous glue to form a multilayered dome. Larvae of the goose-barnacle Lepas use the rapidly hardening secretion of their "cement glands" to attach themselves to ledges of rock when they adopt a sessile mode of life. The bond remains unbroken throughout their lives. Some worms, which live in the sandy floor of the sea, line their tubes with a secretion which hardens soon after it is produced, cementing the adjacent sand grains together to form a solid tunnel wall. The polychaete worm Arenicola builds in this manner. Other polychaetes, the terebellomorphs, sabellariids, and sabellids (10), build free-standing tubes with walls consisting of a membranous, rubbery secretion in which foreign bodies (sand grains, clamshells, etc.) are firmly embedded. Finally, the serpulids (11) build their tubes entirely of glandular secretions. On the peristomium, which is thickened like a collar, there is a paired gland which produces fine granules of aragonite. These are immediately mixed with the secretion of adjacent skin glands and rapidly formed into a tube or added to the existing tube to lengthen it. The mixture solidifies in a short time and forms a firm, indissoluble combination. The males of some bird mites cement themselves to the female during copulation. But this matrimonial bond can be broken and hence does not really belong in the category "rigid-permanent."

1.3 Softening and Re-hardening of Material

Rigid connections belonging in the strict sense to this group do not occur in living organisms. However, this technique of attachment is functionally very similar to all those biological processes described as "growing together," fusion, anastomosis, and others. The callus formation during the healing of broken bones is an example, as is the formation of bone itself.

One way in which formation of new bone occurs is by occupying the site of a preformed piece of cartilage. A human tarsal bone is formed approximately as follows, as described by Hertwig: "From the membrane covering the cartilage, processes of connective tissue, rich in cells, grow (along with vessels) into the cartilage, dissolve the matrix, and coalesce in the center. A network of marrow spaces is produced, in the vicinity of which there is a deposition of calcium salts ("provisional calcification"). The marrow spaces expand more and more by destruction of cartilage. Then bone lamellae are produced by cells located at their surface, and these gradually become thicker. The nucleus of bone thus created slowly enlarges, until finally the cartilage is almost completely displaced, only a thin layer remaining on the surface." Ossification proceeds from one, or sometimes from two, bony nuclei. It begins very late, in the first few years following birth. Here, too, within each nucleus of bone there is a process of fusion and anastomosis of bone-forming elements. In the same way, two bone nuclei may grow together. Their elements interpenetrate one another and mat together; the material continues to "fuse" as long as its cells remain capable of growth and movement. After it has solidified the bone nuclei are just as indissolubly and indistinguishably bound to one another as are two pieces of metal, the edges of which have been fused by welding.

1.4 Connection by Anchoring and Interlacing

This special group comprises permanent connections between two parts without the use of bonding materials or special devices. Anchoring, interfolding, interweaving, and tying occur both in nature and in engineering, in a variety of forms.

The roots of the higher plants achieve a firm attachment to the earth by multiple division and branching, so that a large surface area is formed. This branching allows growth in different directions, as well as penetration into cracks in a mechanically firm substrate. Fine processes are formed which to

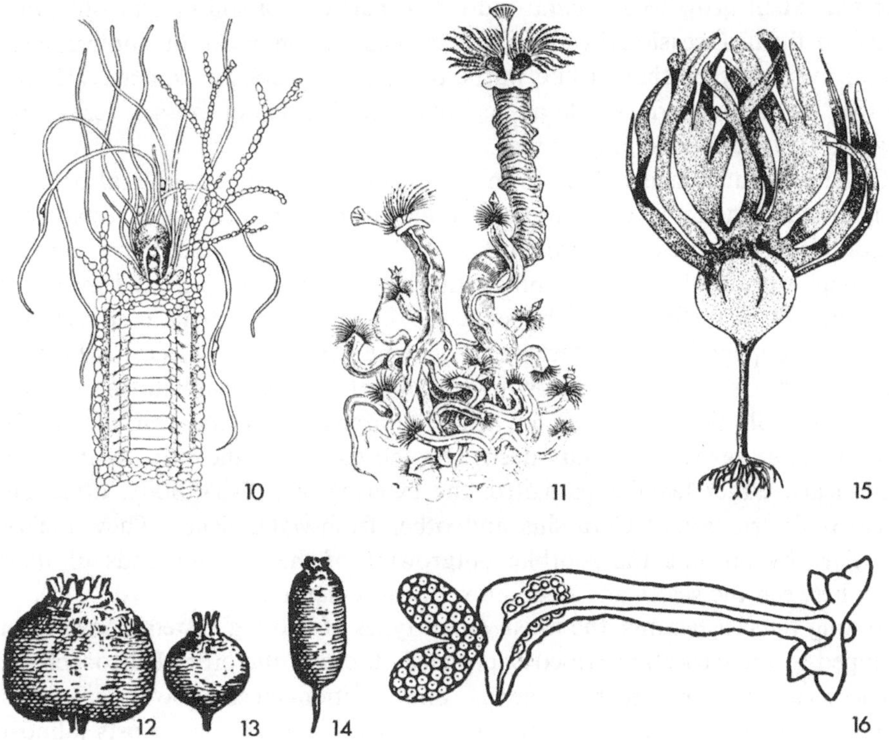

Plate 2 (Figs. 10–16)

10. Sabellidae: Sabella pavonina
11. Some Serpulidae
12–14. Various root shapes
15. Rhizoid of a seaweed
16. Lernaea cyprinacea

a certain extent grow to surround individual particles of soil or can otherwise adhere to them. Occasionally there are thickenings on roots or subterranean shoots; these may be shaped like warts, or irregular knobs or spheres (12–14), and they have a secondary role as anchors. The rhizoids of mosses, ferns, and seaweeds (15) are functionally similar.

Organs of attachment in the form of penetrating horns or roots also occur in the animal kingdom; an example is found in the barnacle-like ectoparasites (Chelonibia) which use the external ridges of their housings, which have rootlike branchings toward the base, to break through the horny layer and even the bony armor of the turtles on which they settle. Massive root systems are developed by parasitic copepods. The anterior end of some female Lernaea (16) is button-shaped and bears hornlike outgrowths with which this ectoparasite can anchor itself in the gill tissue of the cod, where it remains for the rest of its life. These processes grow like roots deep into the tissue, all the way to the wall of the heart. Other females parasitize the exterior of a fish's body, attacking the carps Cyprinus and Carassius and other fresh-water fishes. They anchor themselves by pushing the rootlike outgrowths of the anterior ends of their bodies between the scales and deep into the musculature.

Among the crustaceans, the most varied types of rooting systems have been developed by the parasitic cirripedes, primarily those of the order Rhizocephala.

These root systems are both genus- and position-specific; some examples are given in Figs. 17 to 21. The animals do not live outside their hosts (almost all decapod crabs), but rather are endoparasites. For example, Sacculina carcini (22) is parasitic on crabs: the final larval stage settles on a young crab and rather audaciously, according to Kaestner, "transforms itself into a creature consisting only of an encased system of undifferentiated cells and an injection cannula through which the mass of cells eventually migrates into the crab. There it forms a characteristic rootlike network of tubes for the uptake of food (22). This network, the Sacculina interna *si*, surrounds in particular the gut and muscle bundles of the host crab. After some months the interna breaks through the abdominal body wall of the host and forms a sac, the so-called Sacculina externa (*sa*, 22), which contains the gonads. From the eggs, the free-swimming larval stages develop, the last of which again infests a crab, completing the life cycle of this very peculiar parasite." Biomechanically, part of the Sacculina interna acts as an anchoring device for the externa, which is often a massive protuberance exerting considerable drag upon the point of attachment as the crab moves across the ocean floor.

Root organs which serve for both anchorage and food intake are also found in the larvae of some aquatic mites. The Hydrarachna larva is parasitic on aquatic beetles and bugs. It anchors itself to the legs of the host with a network of roots consisting of chitinous canals (23) formed by evaginated hypodermal cells. Other mite larvae anchor themselves on Anopheles mosquitoes or dragonflies by developing a club-shaped or tubular root organ (24, 25).

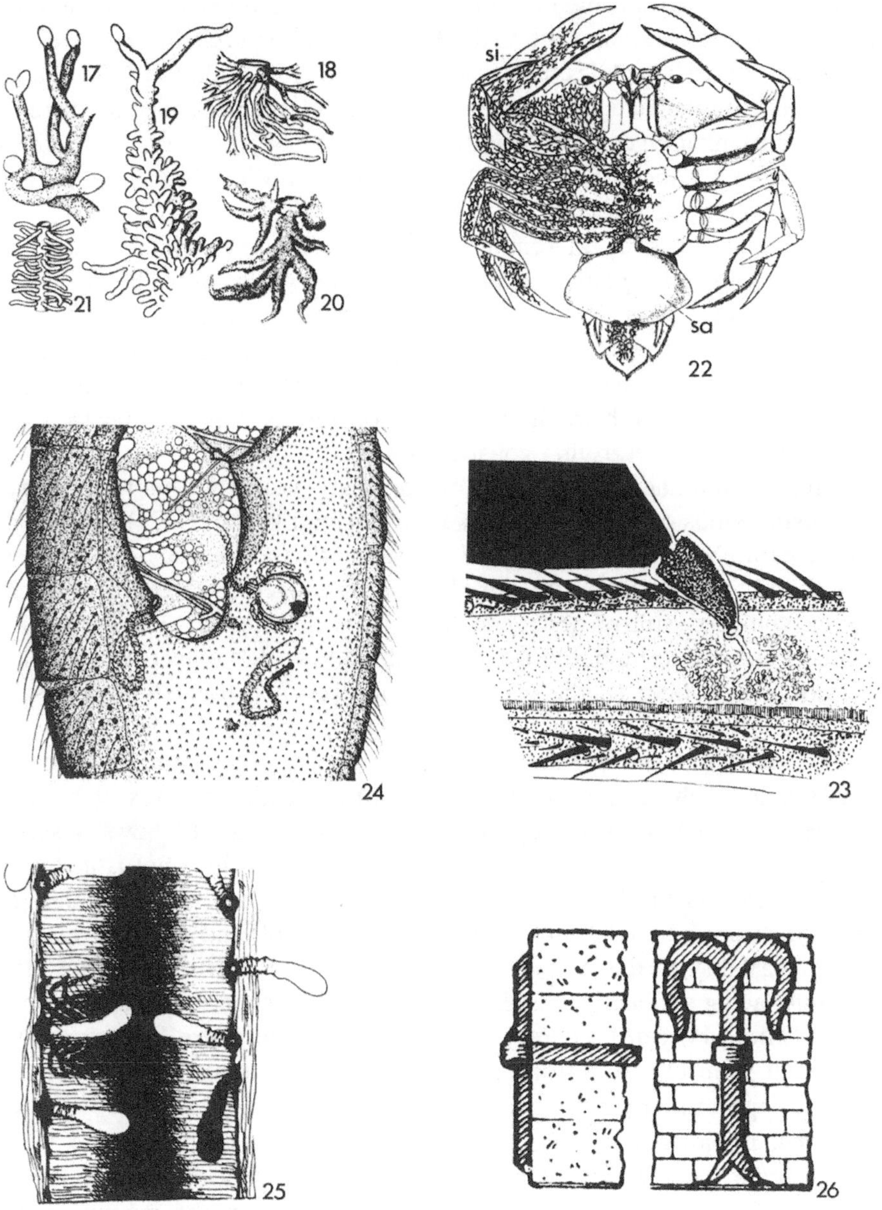

Plate 3 (Figs. 17–26)

17 and 18. Root system of Peltogaster
19 and 20. Root system of Septosaccus
21. Root system of Gemmosaccus
22. Anchoring of Sacculina carcini on Carcinus maenas
23. Anchoring of Hydrarachna in the leg of Notonecta
24. Mite larvae on Anopheles
25. Mite larvae on a dragonfly
26. Anchor plate for tie rod

When subjected to strong mechanical forces, root organs are sometimes inadequate to maintain attachment. In the 3-cm-long parasitic cirripede Analesma, the protective calcareous plates have regressed entirely in evolution. The animal protects itself from damage, and avoids presenting a large resistance to currents of water, by burying itself completely in the skin of its host, the shark Etmopterus spinax. As additional anchorage, the stalk end sends threadlike, branching projections like roots into the musculature of the host.

In engineering an identical principle is used to reinforce attachments: the ends of wall hooks, railing supports and traffic signs which are to be cemented in place are divided in a "rootlike" manner; plates are embedded in sandy or boggy ground to provide jacking pads and anchors; anchor plates are used as attachment points for suspension-bridge cables and for the tie-rods reinforcing the walls of old stone or brick houses (26); concrete is pressure-injected into perforated pipes for underground construction; and so on.

Examples of interlacing and tying of cords are numerous in technology; there are many kinds of weaves and linkages, nets and knots (27–31).

Among nature's weavers certain birds, particularly the weaver birds (Ploceidae), are skilful at intertwining threads. They weave together single plant fibers to make a compact, sturdy nest. The first fibers are tied to a branch with a specialized looping technique. Figs. 32–34 show how the weaver bird Quelea quelea adds a new fiber to the wickerwork already constructed. Figs. 35–37 show some of the kinds of knots used by Quelea to tie the fibers around a twig. The knotting technique used here is reminiscent of a technique formerly used in weaving, where the eye with which the heddle receives the warp thread (the "branch") is formed by two simple knots. Fig. 38 shows a series of man-made knots, used for tying ropes to beams. The Quelea knot is included in some of them. As the bird ties this knot, beak and feet work together in a cleverly synchronized way (39–44). Another commendable and also very durable example of basketry is the nest of the penduline tit Remiz pendulinus (45), in which interlacing of fibers – most likely developed by chance – in the manner of a square knot (46) can be demonstrated. All these nests are hanging nests. They are built from top to bottom, and the first step is to anchor the starting fibers to the supporting branch.

A particularly interesting case of complex anchoring is that of the teeth in the vertebrate jaw (47, 48). In primitive vertebrates they are rooted firmly in place by a fusion of the teeth and jawbone, whereas in the mammals the teeth are resiliently suspended in sockets (or alveoli) in the jaw. Between the root of the tooth and the periosteum of the alveolus there lies a connecting root membrane, the periodontal membrane. It consists of collagen fibers (Sharpey's fibers) accompanied by fibrocytes. The direction in which the fibers run corresponds to the mechanical stresses upon the tooth. Between closely adjacent teeth, such as the incisors of humans, fibers run between the alveoli from the cement at the neck of one tooth to that of the next. The fiber system of the periodontal

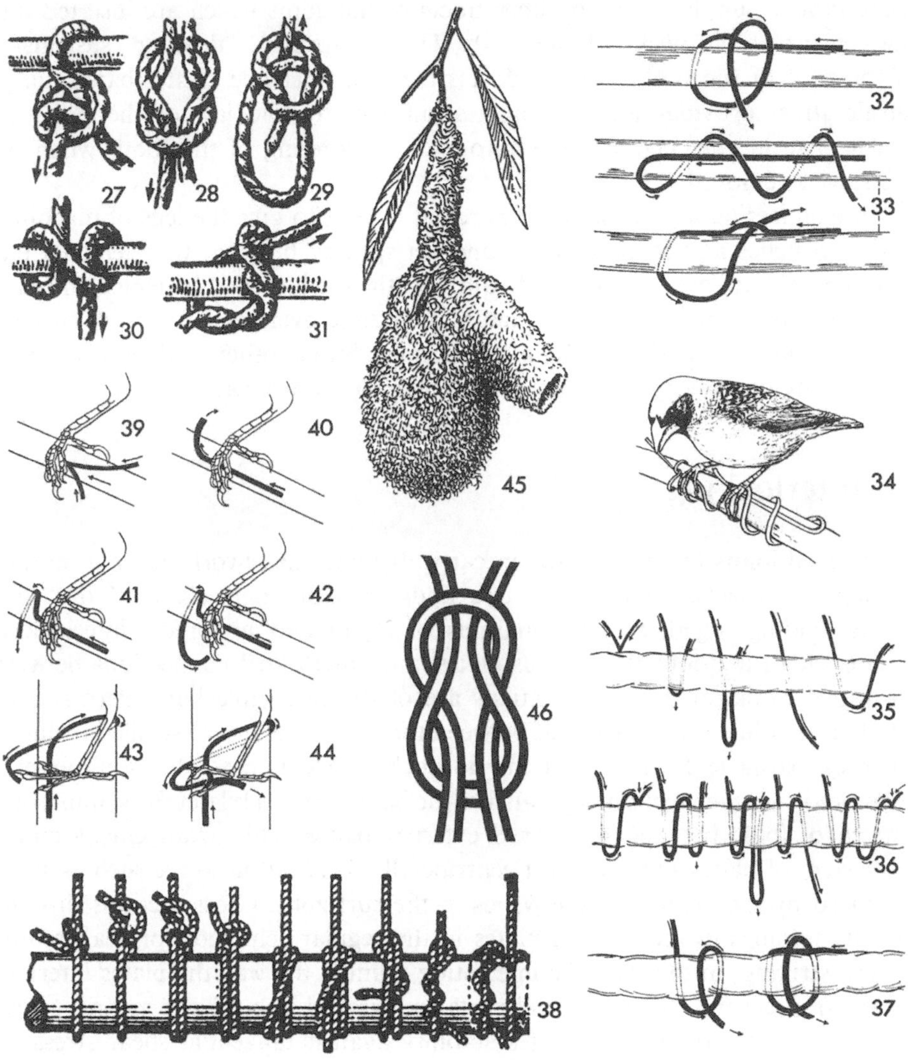

Plate 4 (Figs. 27–46)

27–31. Types of knots
32–34. Weaving technique of Quelea quelea
35–37. Knotting technique of Quelea quelea
38. Types of knots
39–44. Coordination during knot-tying by Quelea
45. Nest of Remiz pendulinus
46. Square knot

membrane is supplemented by fiber tracts of the gums which are inserted into the cement at the necks of the teeth. Thus a springy, "floating" suspension of the root of the tooth is formed; a system of fiber-like elastic bands under tensile stress provides for spacing, transmission to the jaw of the pressures which arise in chewing, and the automatic recentering of the tooth when the pressure is removed.

In engineering a quite similar approach is used to give the legs of machines, optical benches and so on a stable and yet springy bearing; the elements that take the pressure are set into hard-rubber fittings which are cemented in place. Because of their molecular fine structure these take over the function of spacing, force transmission, and damping of oscillations. Still another analogous example is the fixing of screws in walls by means of expansion plugs.

1.5 Interlocking

Interlocked joints like those used in carpentry and metalwork occur in nature. Among the insects, for example, one finds median interlocking of the wing covers, folding together of the margins of elytra and abdomen, interlocking of thoracic or abdominal chitinous plates, and interfolded connections between the wings. Some of these connections are of the releasable kind. Permanently interlocked joints are to be found between the skeletal plates of sessile crustacea, the rock barnacle Balanus and the goose barnacle Lepas. The housing of a balanomorph, when sectioned parallel to the base, shows a bilaterally symmetrical pattern of folds between 4, 6, or 8 external plates. This arrangement makes the system of plates stable against centripetally directed pressures such as those produced by the beating of the waves in the surf zone. Cross sections through the interfoldings of the rows of plates in the regular echinoids, or sea urchins, are less striking; in this case the interesting point is the way the plates interlock as seen from above. The individual parts are arrayed in a zig-zag manner which gives the whole structure of plates stability against tangential shear stress. In technology, various ways of interfolding and interlocking tin and wooden members are used; the most common of these are diagrammed in Figs. 49–58.

Square joints (59), miter joints (60), dovetail joints (61), mortise and tenon (62), and rabbeting (63) are classical techniques in carpentry which were common even in ancient Egypt. In the following discussion, all these kinds of connection will be considered as interlocked joints in the broadest sense.

1.5.1 Simple Joints

The square joint has great disadvantages as far as stability is concerned. Accordingly, it is rare in both engineering (59) and in nature for plates to be joined end-to-end or at right angles without some sort of groove or rabbet. An auxiliary

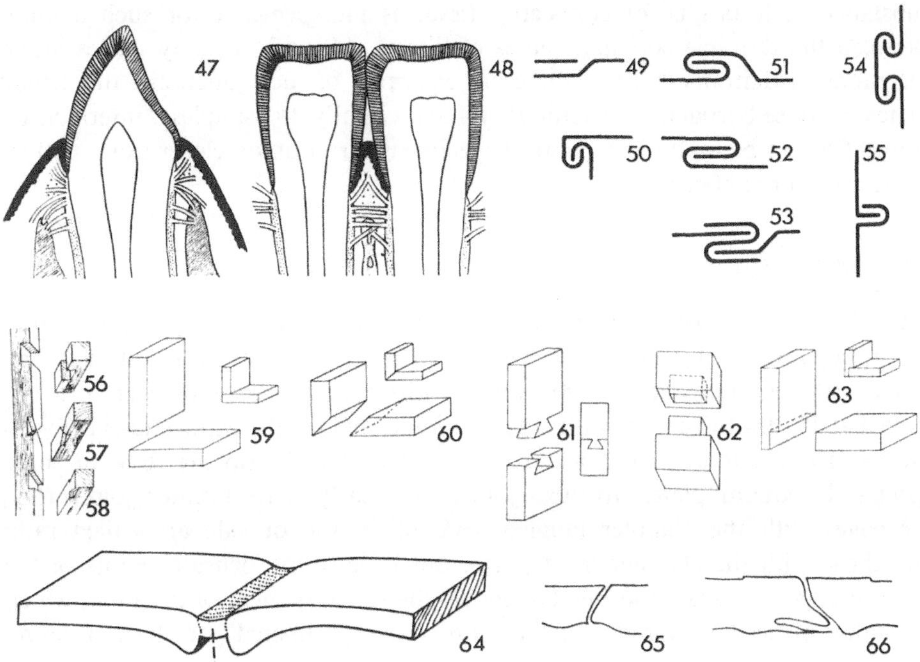

Plate 5 (Figs. 47–66)

47. Human teeth embedded in jaw, sagittal section
48. Human teeth and jaw, frontal section
49–55. Joints used in metalworking
56–58. Joints used in carpentry
59. Square joint
60. Miter joint
61. Single dovetail joint
62. Mortise-and-tenon joint
63. Rabbeted joint
64. Chitin plates of an insect, diagram
65 and 66. Raphe of a diatom

substance such as glue or connective tissue is indispensable for such a joint. Between the thoracic cuticular plates of insects (64) and in many places along the raphe of diatoms (65, 66), there are examples of such junctions; the armor plates of goose barnacles are similarly joined. Usually, the simplest "interlocked" connections to be found in biological specimens turn out, on closer examination, to be miter or rabbet joints.

1.5.2 Miter Joints

The edges of two pieces of wood to be joined are planed on a slant so that when they are put together the pieces are at the required angle to one another. Usually the desired angle is 90°, and in this case the edge of each piece is angled at 45°. Biological examples of miter joints include the connections between the basal and lateral plates of barnacles (Balanidae, Cirripedia; 67). In some species the lateral plates are also joined by nearly perfect miter joints (68); the edges with the chamfer (angled surface) on the outside are called radii and those with the chamfer inside, alae. But usually the delicate points of the mitered edges are seated in fine flanges on the other element of the joint, which amounts to the development of the more stable "rabbeted" or "halved" joint.

1.5.3 Rabbet Joints

In schematic diagrams the joints between the margins of balanid lateral plates are sometimes represented as typical halved joints (69) like those used extensively in carpentry; they are equivalent to opposed rabbets (63). But this common diagram is somewhat over-simplified. In Stephanolepas muricata, for example, almost every plate has strong flanges on the outer surface, but on only one of the two edges. The edge on the other side of the plate is seated beneath the flange of the adjacent plate, more in the manner of a miter joint (70). The flanges are secondary formations. This arrangement permits a very firm interlocking of the barnacle's housing, though it is not ideal. However, balanids often form tightly-packed groups, in which the sides of individuals are flattened (71), so as to withstand the force of the waves as aggregates rather than in isolation. Whatever flexibility exists in such inter-plate couplings may actually enable this flattening. Balanus improvisus (72) comes very close to the mechanically desirable halved joint, at least in the joints between the carina *c* and the carinolateralia *cl;* the ends of the beveled edges of each plate are seated in the ridges of the adjacent plate. A secondary interlocking is provided by the fitting of toothlike projections on one plate into corresponding hollows on the other. This principle of simultaneous coarse and fine interlocking is common, and will be noted again in the catch between the elytra of beetles (79).

The details of the interlocking arrangements used in the joints between barnacle plates depend on the one hand upon the level of organization of

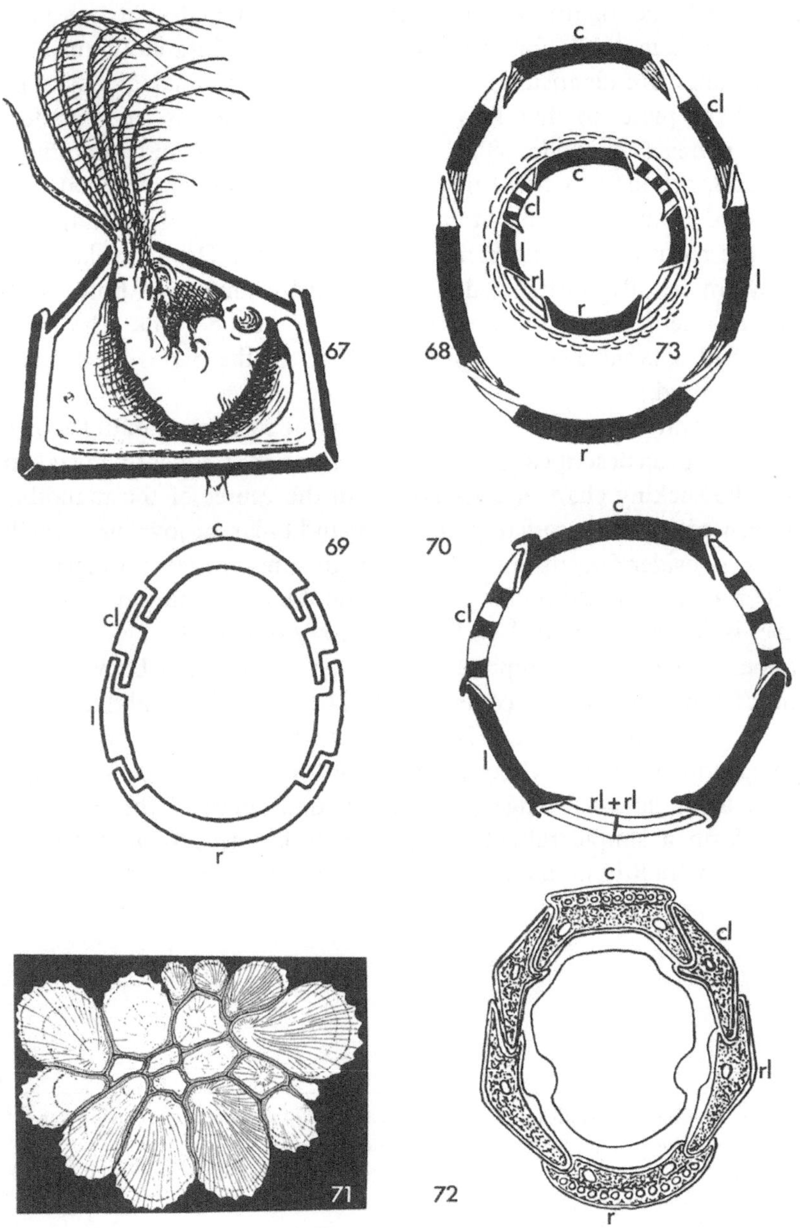

Plate 6 (Figs. 67–73)

67. Miter joints between basal and lateral plates of balanids
68. Miter joints between lateral plates of balanids
69. Halved joint in a balanid
70. Flanged joints in Stephanolepas muricata
71. Compressed aggregation of balanids
72. Modified halved joint in Balanus improvisus
73. Alternating miter joints of Cataphragmus

the genus concerned and on the other, upon the pair of plates in question. In the primitive Chthamalidae, for example in the genus Cataphragmus, the carina *c* and rostrum *r* are identical in shape and have outward-facing chamfers on both sides of the plate, so that they are covered by the inward chamfers of the adjacent plates (73, inside 68). The carinolateralia *cl* and lateralia *l*, on the other hand, are asymmetrical in primitive genera. On the carinal side they end in inward chamfers which fit the outward chamfers of the adjacent plates, and on the rostral side the arrangement is reversed. The rostrolateralia *rl*, finally, are symmetrically constructed, with outward chamfers on both sides. Fusion can occur, altering both the number and type of the joints. The most extensively modified balanid genus is Pyrgoma; here all the plates have fused to form a solid pyramid.

The larva of the water beetle Dytiscus has powerful sucking mandibles (*md*, 74). These have often been described, but the descriptions are frequently mistaken in one respect: the sucking channel does not lie in the center of the mandible but rather is formed by a marginal fold, the mandibular groove *mr*, which is closed over a considerable distance by a tongue-and-groove arrangement (75). Fine teeth provide an additional means of hooking the margin together, and the channel is completely sealed off by dense rows of hairs *h*.

Simple rabbet joints are also represented by the connections between the outer margins of the elytra (wing covers) and the sides of the abdomen of many beetles. In the aquatic beetle Hydroporus ferrugineus the elytra bear a trace of a longitudinal groove, while the edges of the abdomen are projected upward to form a longitudinal ridge (76). The medial edges of the elytra of this beetle also form a simple rabbet joint, at least in the proximal region, where they are seated tightly in lateral infoldings of the scutellum (77).

Plate 7 (Figs. 74–83) ▶

74. Head of a larva of Dytiscus marginalis
75. Cross section through the sucking mandible of the Dytiscus larva
76. Interlocking of abdomen and elytra in Hydroporus ferrugineus. e_l, e_r, left and right elytra; ee, elytral ridge; hw, hindwing; ple, pleural ridge; te, tergum; st, sternum
77. Interlocking of scutellum and elytra in Hydroporus ferrugineus; labels as in (76)
78. Double tongue-and-groove joint between the elytra in Hydroporus ferrugineus; labels as in (76); hf, horizontal tongues; vf, vertical tongues
79. Scallopped tongue-and-groove joint in an insect
80. Interlocking within the oral cavity of the Dytiscus larva
81. Detail of the interlocking mechanism of (80)
82. Schematic cross section through a hawk-moth proboscis
83. One pair of elements from the interlocking array of a hawk-moth proboscis

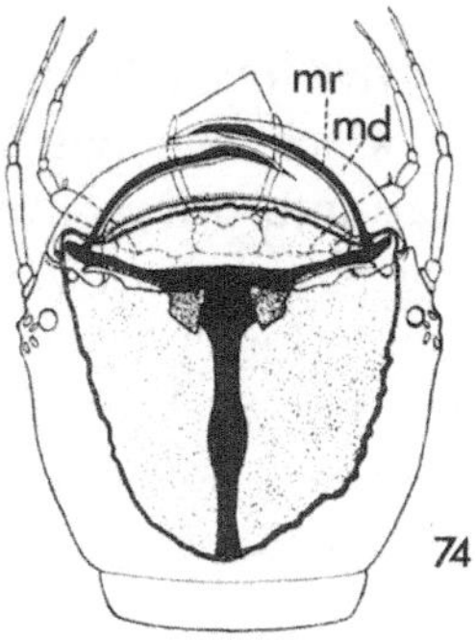

74

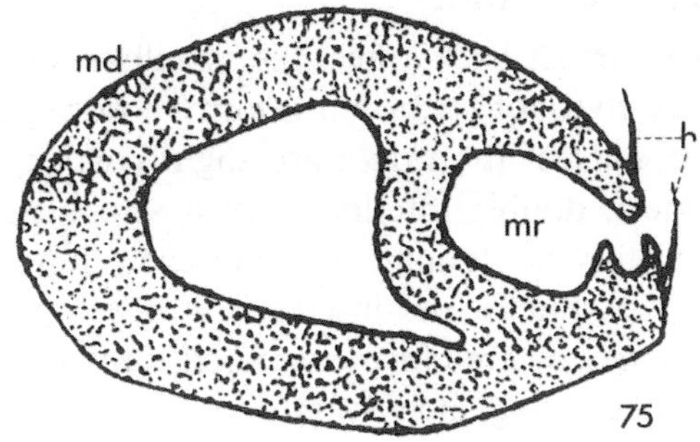

75

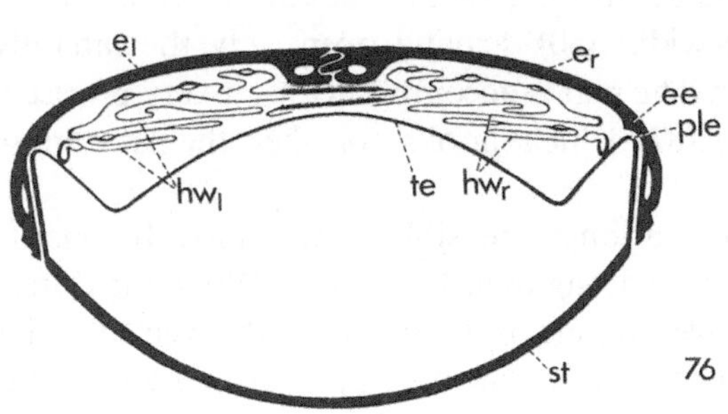

76

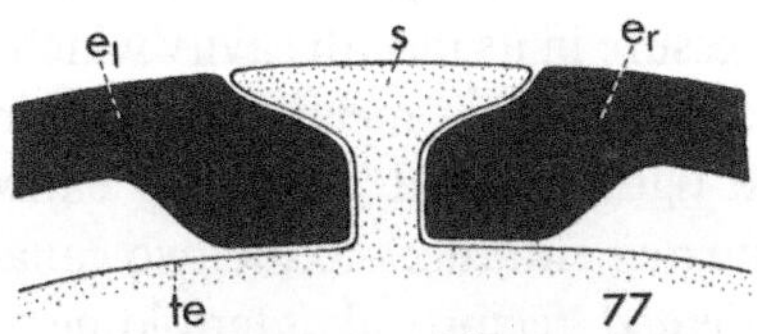

77

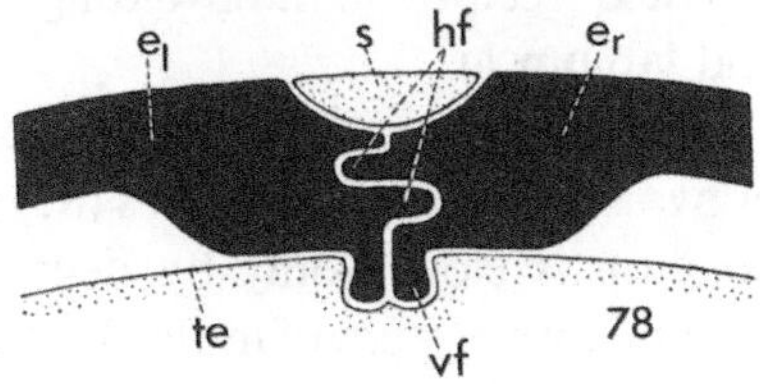

78

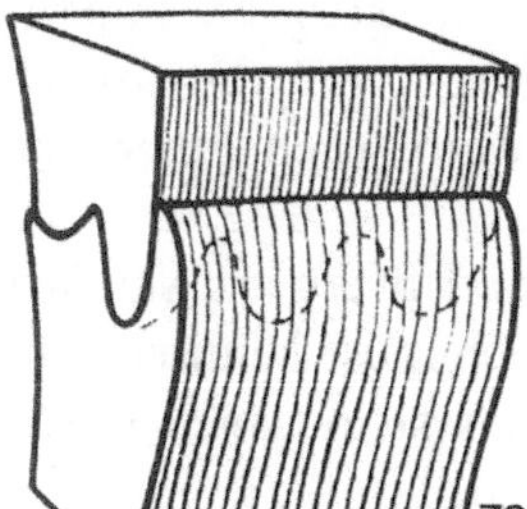
79

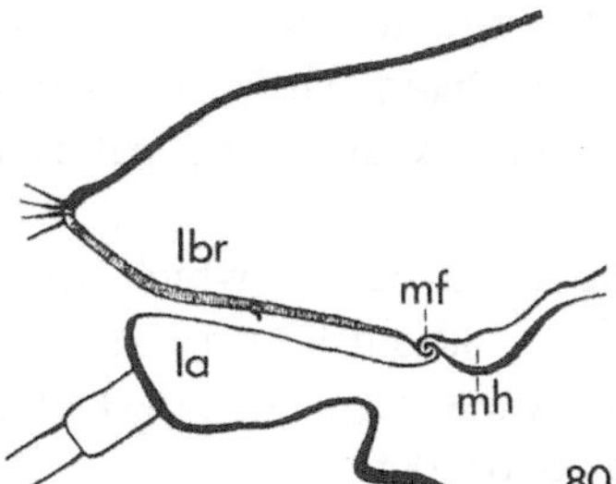

80

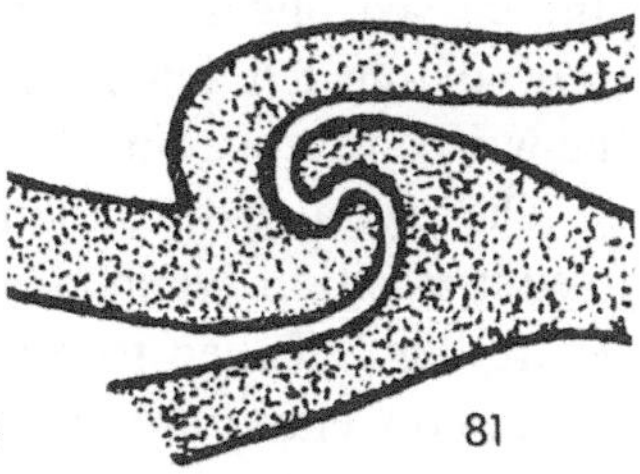
81

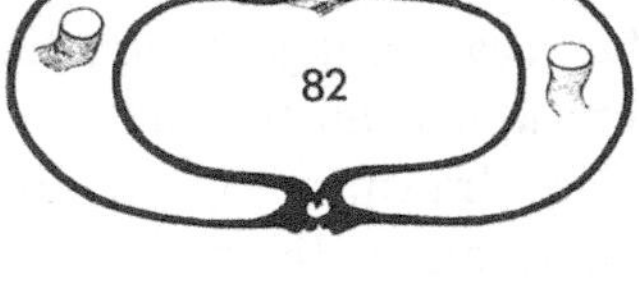
82

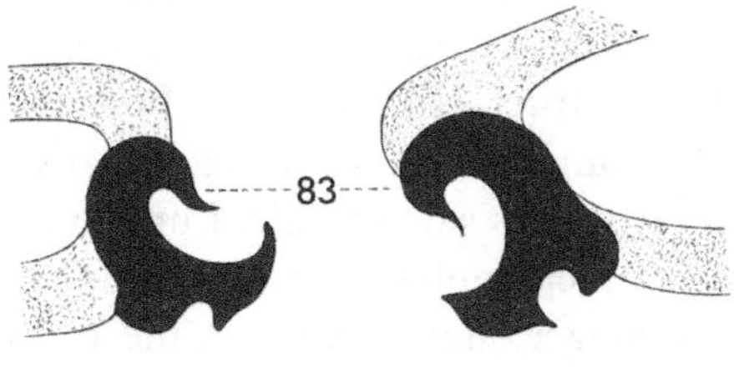
83

1.5.4 Dovetail Joints

Somewhat further back, where the scutellum stops and the elytra touch directly, each of the elytra of Hydroporus forms a ridge and a groove, which interdigitate. Since the ends of the ridges are slightly enlarged, they snap into the grooves to provide a double interlock which strongly resists being pulled apart (78). A trace of this stable joint is found even in some of the more highly evolved balanids (72), but in the insects it is developed optimally. Here such double fastenings may be even further secured if the ridges and grooves are not straight but undulate along their length so as to provide an additional interlocking mechanism as in Fig. 79.

When the Dytiscus larva draws out the body fluids of a captured animal with the sucking mandibles just described, it must produce a lowering of the pressure in its mouth cavity which in turn causes the inflow through the mandibular tubes. This means that the frontal opening of the mouth cavity *mh* must be tightly sealed across its entire width (80), leaving open only the laterally situated entrances to the two canals. The seal is accomplished by a very precisely oriented reciprocal interfolding *mf* (81) which hooks together the labrum *lb* and labium *la*.

The proboscis of the hawk moth is a long extensible suction tube. It consists of two semicylindrical halves (82) joined along their long edges. While the dorsal junction is closed only by dense overlapping rows of hairs, the ventral joint is made by a system of multiple attachments similar to a zipper. Strong sclerotized J-shaped elements are aligned at an angle opposite one another (83) so that each forms an individual double interlock. The precise arrangement is difficult to describe, since the matched elements are themselves twisted. According to Schoenichen, one can visualize the mechanism to a certain extent by hooking "the curved fingers of the right hand (with the exception of the thumb) into those of the left hand, and then twisting the palms toward each other until the wrists are touching."

The elytra of those lamellicorn beetles which are good flyers are similarly fixed in position, when at rest, by about fifteen clasps which must be disengaged before flight. Taken together, these form an interlocking system which is very like the dovetail joint in carpentry (84).

1.5.5 Mortise Joints

The mortise and tenon form a composite structure useful for joining relatively small surfaces in carpentry. Round mortises, to receive pegs (85), are particularly easy to construct. However, they have the disadvantage that they are not secure against rotational stress. One can compensate for this disadvantage by locating two such devices side by side. Functionally equivalent joints can be made with square or rectangular elements, two types of which are drawn in Figs. 86 and 87. The longer such joints are, the more they resist rotation. In the extreme

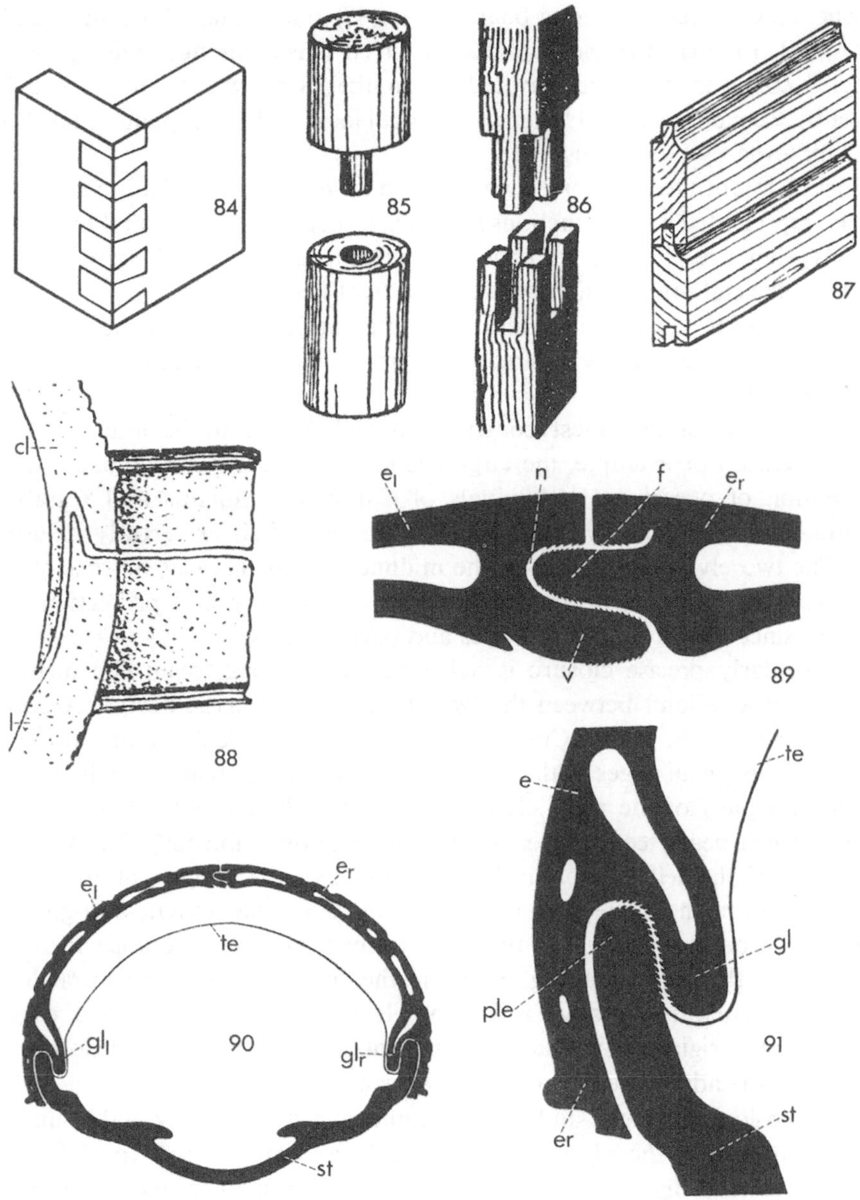

Plate 8 (Figs. 84–91)

84. Dovetail joint
85. Round mortise and tenon
86. Crossed mortise and tenon
87. Tongue-and-groove joint
88. Interlocking of the plates of Stephanolepas
89. Double tongue-and-groove joint between the two elytra in Cnemidotus caesus; labels as in (76)
90 and 91. Glossula-type interlock between elytra and abdomen in Cnemidotus caesus; labels as in (76)

case, where the elements are long boards as in (87), one speaks of a tongue-and-groove joint. In carpentry, such joints are given tensile strength with glue or an angled construction as in dovetailing; in the living world, they are held together by secondary elements such as spherical locking devices, plain or toothed ridges, and even by dovetailing.

The principle of a round mortise with a tenon which is locked in ball-and-socket fashion is represented by the junctions in the leg or between the antennal segments of insects; they are "flexible-permanent" connections and thus belong among the flexible skeletal joints which we shall not treat here.

The miter joint between the lateralia *l* and carinolateralia *cl* of the barnacle Stephanolepas is additionally secured by a narrow but deep tongue-and-groove connection (88).

Mortise joints of the finest construction and fit are to be found in the elytra of beetles. For example, the large beetles of the family Carabidae—with the exception of occasional individuals of Carabus granulatus and Carabus clathratus—are unable to fly, since the membranous wings are greatly reduced in size. The two elytra are united in the midline by a permanent mortise joint. In this way they form a strong, semi-ellipsoidal shield, useful protection for the beetles since they live among stones and particles of soil.

A particularly precise closure is achieved by the long, seam-like median tongue-and-groove joint between the two elytra of the aquatic beetles. A good example occurs in the beetle Cnemidotus caesus (Dytiscidae). At the location of the seam, the dorsal edges of the elytra form a butt joint. Below this, however, a toothed tongue *f* of the right elytron e_r exactly fills a groove *n* in the left elytron e_l which bears teeth angled in the opposite direction (89). The ventral projection *v* of the left elytron is built almost like the tongue of the right elytron and bears the same sort of teeth. Thus a double tongue-and-groove joint seems to be indicated. It is not fully formed, however, because there is only a trace of the corresponding groove in the right elytron; the ventral half of the second enclosing groove is lacking. It is evident that the two edges of this seam are originally symmetrical, and only in the course of development of the imago (the adult insect) does asymmetry arise.

This tooth-locked tongue-and-groove connection is absolutely stable under vertical distortion of the elytra by external forces. An attempt to slide the elytra apart toward the side by way of the normal opening muscles, however, encounters no resistance. The tongue becomes somewhat more narrow distally, and the teeth can be disengaged by suitably shifting the wings.

The interlocking device involving the so-called glossula, which couples the edge of the pleuron on each side to the edge of the corresponding elytron in Cnemidotus (90, 91) is functionally quite similar in structure. Between the margin of the elytron *e* and the glossula *gl* associated with it there is a deep groove into which the edge of the pleuron *ple* fits precisely. Oppositely directed "sawteeth" on the contacting surfaces of pleuron and glossula again form an additional locking device. Altogether, the elytra are tightly wedged in position by three clasp mechanisms—with each other, with the scutellum, and with the abdomen—as well as by two additional seating mechanisms in the wing joint and in the so-called "catastigma." According to Heberdey, the process of opening the elytra obviously requires quite a complicated coordination of movements, while the closing process occurs almost automatically.

In other dytiscids, as well as in almost all other families of the aquatic beetles, there are comparable interlocking connections. These families include the Gyrinidae, Hygrobiidae, Haliplidae, and Dryopidae. It is of vital importance to these insects that the elytra never be opened "inadvertently" or by external forces when the animal is underwater. Beneath the elytra the eight large pairs of spiracles of the tracheal system open into the air-filled subelytral space. These must never come into contact with water.

In engineering, similar interlocked connections between elements in a series are of common occurrence. In woodworking, where boards are to be fastened side by side, they are ordinarily joined by halving (92) or by a tongue-and-groove arrangement (87). The tongue of one board is pressed into the groove of a second which, in turn, has on the other side a tongue which fits into the third element, and so on.

One type of roofing tile, a sort of pantile with a "claw" on one edge, is laid so as to form a similar sequence of overlapping joints (93). Other roof tiles, of the so-called "plain single Roman" type, have ridges and hollows on the edges which are fitted together (94). These marginal structures are less marked in the simple pantile (95), where the whole tile is curved into an S-shape. Fish scales are also not uncommonly folded together at the edges. The biological counterparts of pantiles would be the scales of the ganoid fishes. The interlocking and overlapping scales in Amblypterus duvernoyi are shown in Fig. 96 from inside (i) and outside (a).

1.5.6 Multiple Interlocks

In this category we include cases where more than two elements are joined at one point, or where just two elements are connected at several places. An example of the former is given by the barnacle Xenobalanus globicipitis (97), where two wall plates *w* and the flanged end of a third plate *a* are wedged together. There are analogs of this situation in the construction of furniture. The second case, a compound junction, is realized in the barnacle Lithotrya valentiana. This animal bears strong calcareous scales on the capitulum, the margins of which have hard teeth for boring into the substrate. These calcareous scales can apparently interlock by means of about eight processes arranged along the margins (98).

A very precisely developed compound system is found in the water beetle Hydroporus mentioned previously (78). In the region of the scutellum the elytra

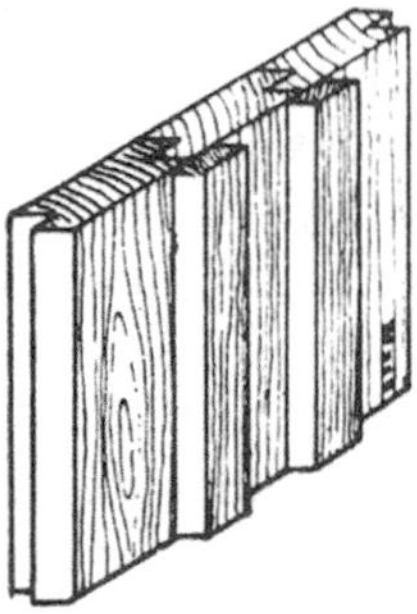

92

93

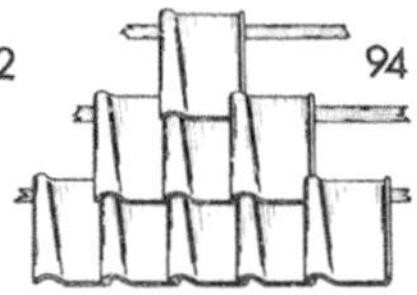

94

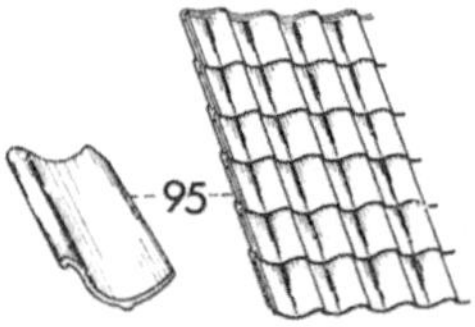

95

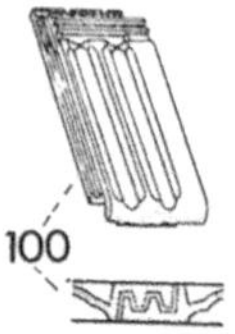

100

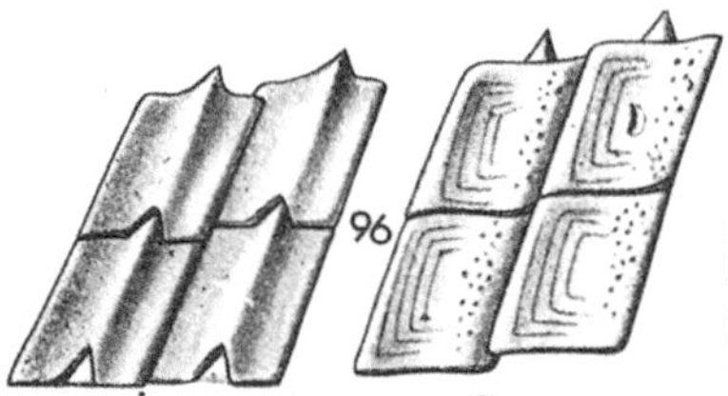

96

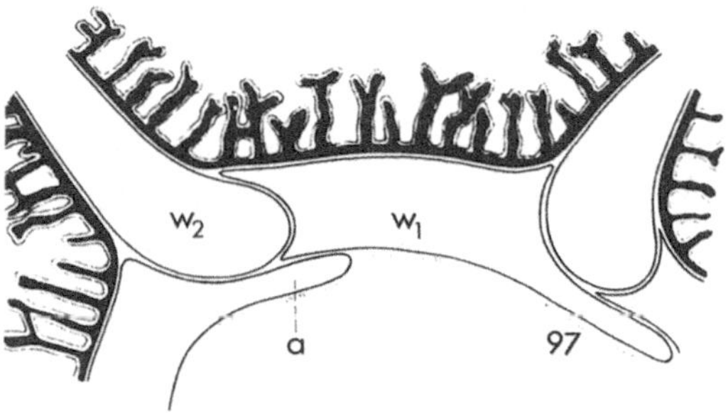

97

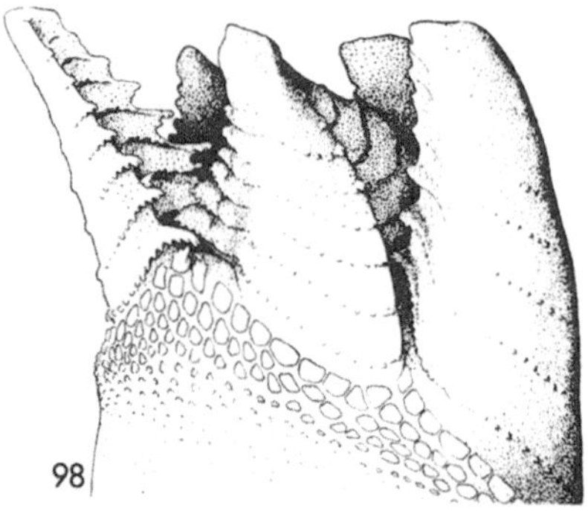

98

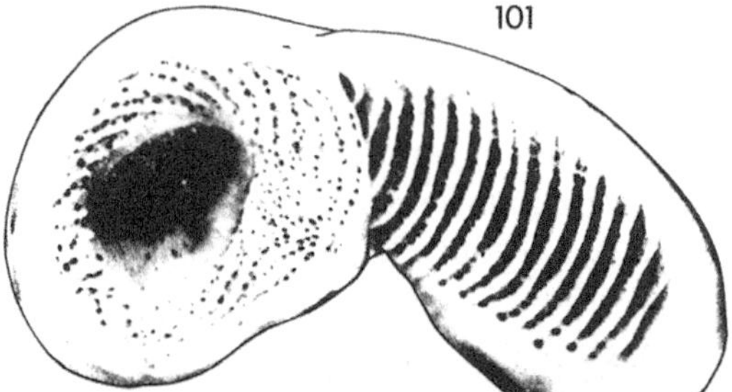

101

103

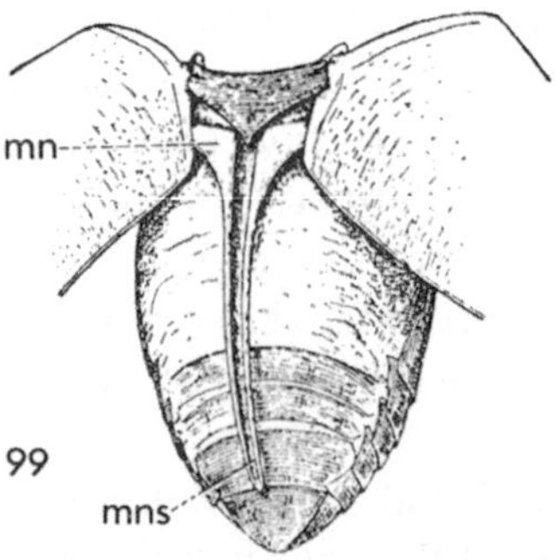

99

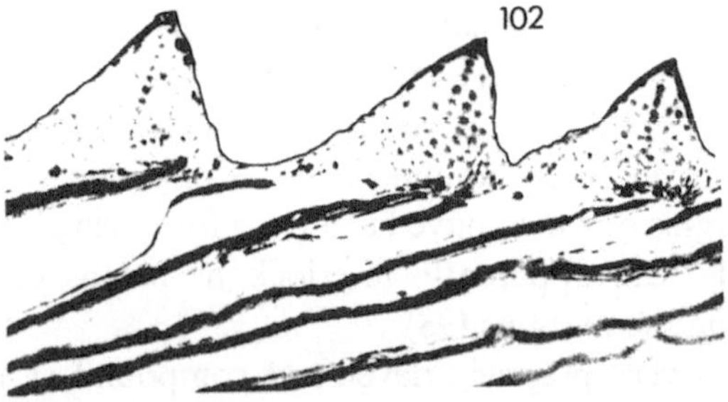

102

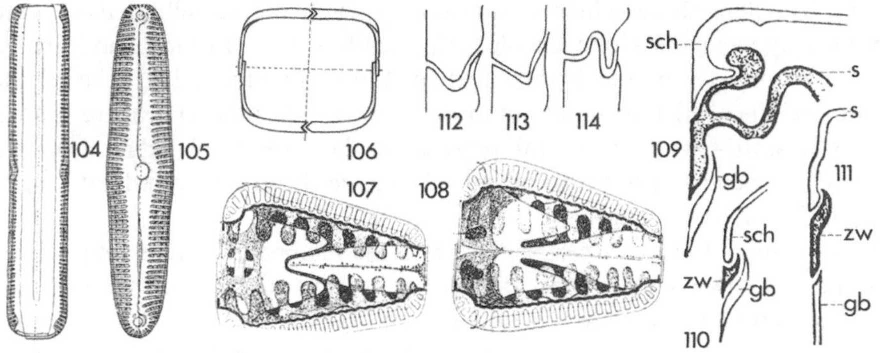

▲
◀ Plate 9 (Figs. 92–114)

92. Halved edge joints
93. "Claw" pantile
94. Plain single Roman tile
95. Pantile
96. Interlocking between scales: joints between rhomboidal ganoid scales, Amblypterus duvernoyi (Heterocerci)
97. Mutual interlocking of three plates in Xenobalanus globicipitis
98. Multiple interlock in Lithotrya valentiana
99. Subsutural-metanotal interlock in Pholeuonidius pinkeri
100. Flat tile interlocking at head and sides
101. Postopercular organ in Kneria
102. Longitudinal section through (101)
103. Diagram of the "jig" mechanism of (101)
104. Side view of Pinnularia viridis
105. Pinnularia viridis, from above
106. Pinnularia viridis, cross section
107 and 108. Formation of hypovalves in Surirella calcarata
109. Girdle joining the valves of Grammatophora marina. gb, girdle; s, shell band; sch, shell; zw, intercalary band.
110. Girdle in Epithemia turgida; labels as in (109)
111. Girdle in Climacosphenia moniligera; labels as in (109)
112–114. Raphe of Pinnularia maior, sectioned at different levels

have no less than three different attachments: with the scutellum itself, with each other by means of a remarkable reciprocal tongue and groove, and finally, with the tergum *te*. In the broad, median dorsal longitudinal groove of the tergum, the so-called metanotal suture (cf. *mns* in 99), both elytra are seated, each by a sturdy median ventral process *vf*, the subsutural ridge. The two ridges are pressed against one another and together form a longitudinal tongue which fits tightly into the groove.

This two-part tongue is perpendicular to the reciprocal tongue and groove connecting the two elytra. Since the strong, rigid scutellum is situated just above this compound junction, it ought not to be possible to release this firm closure if the scutellum were fixed vertically with respect to the metanotum. Before the elytra can be pulled apart, the subsutural-metanotal interlock must be disengaged. This can only happen if the metanotum is pulled downward. Thus the first step in the coordinated release of the elytra is a contraction of the dorsoventral muscles in the metanotum. Since the sterna (the ventral plates) are in general stronger than the nota (the dorsal plates), it is primarily the nota which move and by their downward motion release this "safety catch."

Among the Silphidae (carrion beetles) there are certain Bathysciinae (99) in which the metanotum *mn* is continued into a remarkably long, caudally directed projection, along which the metanotal suture *mns* continues as well. The subsutural ridges on the elytra are also much lengthened, so that the interlock extends almost along the entire median border of the elytra. Such an extensive fastening may be necessitated by the strong mechanical stresses associated with the animal's habit of burrowing in carrion. One could also imagine that the groove need not run on such an odd projection of the thorax, covering the abdomen, but rather directly in the terga of the abdomen itself. This case, too, is realized. In some representatives of the Anthribidae and in one genus of the Anthicidae the last abdominal tergite, or the last but one, possesses the necessary longitudinal suture.

Multiple interlocks of this sort have also been devised for roofing tiles. The flat interlocking tiles shown in Fig. 100 have two trough-shaped longitudinal depressions as well as double folds at the head and sides. They are thus doubly secured against displacement both down the roof and across it. Among other forms of roof tiles with a double interlock are the "double Roman" and "Bridgewater" types. Flat tiles with a flange only on the side can be produced "continuously" by cutting up a strip of clay, but if another system of ridges and depressions is to be added perpendicular to this, the tiles must be pressed out individually.

One peculiar wedge structure should be mentioned here. This is found in the males of the African fish genus Kneria. These fish are spindle-shaped, cylindrical in cross section, and live in flowing water. In copulation the partners press close to one another, side by side. Adhesion is brought about by a suction cup on the operculum (gill cover) as well as by the so-called postopercular organ which is next to it, behind the gill slit. This latter organ of adhesion consists of extra large, densely packed scales, one behind the other, which have evolved by fusion of three ordinary scales in a vertical row. The posterior edge of the scale possesses a strongly keratinized epidermal surface which is formed into ridges (101, 102). The middle of the organ is flush with the surface of the body, while the more dorsal parts stick out further and further above the surface. The hard ridges of the organ become caught in the body surface of the female and act as an adapter or "jig" (103) to match the two nearly cylindrical bodies, which otherwise would touch only along a line—a mechanically undesirable situation. In the midst of a flowing stream, such a positioning device for copulation is evidently necessary. The closely related genus Parakneria, which lives on the bottom and presumably also mates there, characteristically lacks this postopercular organ.

Some round pill boxes and rectangular cans have matching ridges and grooves which hold the lid to the box. Ridges in the lid lie opposite grooves in the box, and vice versa, forming a functional unit, so that this connection, in a sense, may be considered as a kind of compound interlock. It is common knowledge that this box-and-lid principle is represented in the connection between the two halves of the siliceous walls of both rounded and angular diatoms. One half, the epivalve, fits like a "cover" onto the other, the hypovalve (104, 105, 106). During asexual reproduction the nucleus and protoplasm of the cell divide and push the two valves apart. Now each half, containing its own nucleus and protoplasm, becomes an epivalve, and the missing hypovalves are formed anew (107, 108). The rims of the valves may be further extended by bands or "girdles" running around or between them. The edges of each such girdle meet in various mitered and interlocking junctions (109–111). During cell division the girdles may enclose the space in which the new hypovalves are being formed.

A peculiar longitudinal fissure in the valves, the so-called raphe, appears in most pennate diatoms and offers an opportunity for the study of all sorts of interlocking connections (112–114). Often the surface of the raphe is indented and twisted as well. The edges may actually touch or may leave open a narrow slit through which plasma may be extruded to effect movement of the animal from place to place. Cross sections at various positions therefore show different kinds of interlocks, ranging from the simple butt joint, through miter and mortise joints, to the reciprocal tongue and groove (114). The raphe is a cleft within an already rigid shell structure, and thus does not join two separate parts. The functional significance of its basically wedge-like construction may therefore be construed as preventing the two parts of the valve, on either side of the raphe, from bulging in when pressure is applied to one side. Quite a few diatoms live between sand grains on the ocean floor and must often be exposed to such stresses.

2. Releasable Attachments

The mechanisms described in this category are designed to permit two different structural components to be quickly coupled or uncoupled—although the two parts are held together firmly as long as the connection is maintained. The most secure of such temporary junctions are those in which one element is pushed into another, by the key-and-lock or plug-and-socket principles. The arrangement of various sliding surfaces can prevent certain displacements of the two parts, so that the resulting degree of freedom of relative motion is strictly limited; the partners can only move with respect to one another along a common longitudinal axis. Various safety catches can provide security against unintentional release of the connection.

This principle presupposes an exact morphological correspondence of *two* parts. Another principle within this category operates with only *one* of the elements being highly specialized, such as in a suction, hook, or gluing device. Coupling elements of this sort may attach to very different structures or animal bodies. No special demands are made of the attachment point, except that in some cases it must not have more than a certain degree of roughness (suction apparatus) or smoothness (hook organs), or that it must be free of mucus, oil, and grease (gluing mechanisms). Thus one may divide this category loosely into two main groups: connection of two complementary parts, and attachment by one specialized device.

2.1 Connection of Two Complementary Parts

2.1.1 Plug and Socket

The two principles characterizing this category—limitation of movement to one direction of translation and, even in that direction, insurance against inadvertent displacement—are most simply illustrated by the modern German electrical plug and its matching socket (120, 121). The plug may easily be pushed into the socket; rotational misalignment is prevented by two opposed guides *f* in the plug which fit into two corresponding grooves *n* in the socket. Since the fit of the guides is to a closer tolerance than that of the prongs in their spring contacts, the prongs and contacts are protected from excessive shear forces. Tilting movements are prevented by the fact that the cylindrical housing of the plug *r* is seated in the recessed, correspondingly shaped receptacle. Two spring ground (or "earth") contacts *c* in the socket bear on two corresponding contact strips at top and bottom of the plug which usually are bent slightly away from the housing in front. These snap the plug into place and hold against a moderate pull. Where a plug is to be connected to a free extension cord, either it or the socket often has a strong metal clip which snaps over a detent into a groove on the complementary part and thus secures the connection.

In the animal kingdom, examples of such temporary, rigid plug-and-socket connections are the deeply undulating edges of the shells of large clams (Tridacna, 177) and the teeth of the crocodile, which occasionally fit into groove-like lateral cavities in the opposite jaw. Some arthropod joints include paired pegs, shaped so as to fit exactly into recesses in the adjacent part, which prevent the joint from twisting when in the closed position. Extraordinarily finely matched plug-type connections are found in the male and female genitalia of a number of arthropods. In particular, insects that copulate during flight (such as the midges and dragonflies) have very precisely fitted parts and a variety of security devices.

The problem of bringing the very fine openings of the male and female genital ducts into absolutely reliable contact is crucial for the maintenance of the species. It is not at all easy to solve. Copulation often lasts for only a short time, and the partners are usually flying in an irregular manner while their abdomens, which may be curved and twisted, vibrate vigorously. The copulating animals cannot control visually the coupling of the genitalia, which are usually located at the extreme end of the body, but rather depend upon information from the sense of touch and proprioception.

The mating devices must therefore be built such that a "hard dock", with the elements properly aligned, is practically obligatory—even though the distal parts of the copulatory apparatus may not initially lie in the optimal orientation. Furthermore, as long as copulation is in progress this contact must be sufficiently secured that stretching and violent oscillations cannot break it. That is, it must fulfill the somewhat contradictory requirements of automatic locking in, very precise fit, and reliable resistance to tensile stress. One is immediately reminded of the in-flight refueling of airplanes and the docking of spacecraft in orbit; in fact the technological design problems correspond closely to those which have been solved by natural selection.

The midge Tanytarsus sylvaticus (Chironomidae) copulates, on the average, for 30 seconds. Copulation is often begun in flight and then continued upon a substrate. In the female, the distal abdominal segments, beginning with the eighth segment, have been modified into a copulatory organ. A medial indentation in the anal margin of the eighth sternite (VII *st*, Fig. 115), the subgenital sinus, is continued cephalad as the hollow genital chamber *gch*, at the base of which lies the unpaired opening of the spermatheca. The chamber is closed off at each side by a genital lobe *gl* densely covered with bristles. Laterally, posterior to these, are the paired lateral lobes *ll*, and the whole apparatus is completed by the paired cerci *ce*. Dorsally almost the entire copulatory apparatus is covered

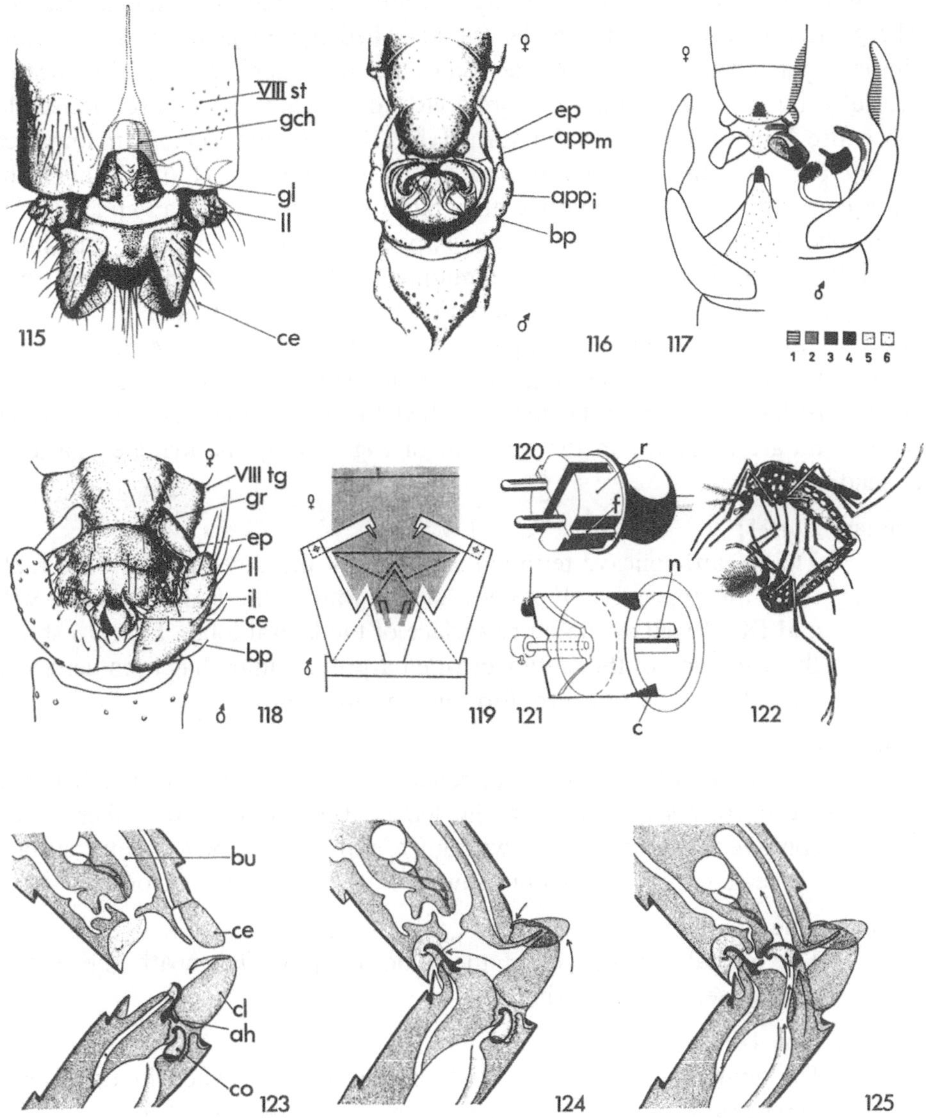

Plate 10 (Figs. 115–125)

115. Genital segments of Tanytarsus sylvaticus ♀
116. Copulation in Tanytarsus sylvaticus
117. Schematic of the principles of operation of (116)
118. Copulation in Limnophyes pusillus
119. Schematic of the principles of operation of (118)

120 and 121. German electrical plug and part of its socket

122. Aëdes aegypti in copulation; circled region is enlarged in (123–125)

123–125. Approach to the copulatory position in Aëdes aegypti

by a shield formed from the helmet-like tergite of the ninth abdominal segment. The genital apparatus of the male, the so-called hypopygium, must therefore slip into place from the ventral side. Since the male genitalia are also located ventrally, but the male copulates from a face-to-back position above the female, the entire hypopygium must be turned by 180°. It consists of an unpaired central part, the anal projection, and paired lateral parts (116). Each of the latter consists of an outer pincer with a basal segment *bp* and an end segment *ep*, a medial appendage with a hook-shaped element app_m, and a thin inner appendage app_i, looped toward the midline of the animal and ending in a sort of tassel. Functionally, the unpaired central part is equivalent to the prong of a plug and the paired outer parts, to safety catches. There are no less than 6 contact zones between the male and female genitalia, for each of which the complementary parts are delicately matched to one another. The parts which correspond are indicated in the diagram of Fig. 117 by the matched patterns of shading:

Contact zone 1:

The inward-concave terminal segments of the outer palps of the male clasp the appropriately convex lateral wall of the female tergites VIII and IX, as well as the outer surface of the lateral lobes. Fig. 116 shows the beginning of the embrace; when a proper union has been achieved, *bp* and *ep* point outward, forming an obtuse angle.

Contact zone 2:

The hooks of the medial appendages of the male are inserted into the narrow space between the female cerci and lateral lobes. Their concave inner surface lies without a gap on the convex outer surface of the cerci. Fig. 116 again shows a state of incomplete closure.

Contact zone 3:

The medial appendages of the male are pushed beneath the female cerci to serve as a bearing.

Contact zone 4:

The anal projection of the male hypopygium is seated in the female subgenital sinus, where it bears upon the two genital lobes.

Contact zone 5:

The curved inner appendages of the male, with their laterally directed terminal tassels, are pressed into the medial concave surface of the female cerci.

Contact zone 6:

The penis of the male, which is below the anal projection, is pushed into the subgenital sinus of the female; its tip reaches the unpaired opening of the ducts leading into the spermatheca, completing the connection which is necessary for transfer of the sperm.

The first five contact zones listed guarantee an exact positioning of the crucial sixth zone (the prong of the plug) and ensure that the entire arrangement is not pulled apart (zones 1 and 2), bent (zones 1, 2, 3 and 4), or rotated about the long axis (zones 1, 2 and 3). The delicate contact zone 5 presumably has no particular mechanical function but rather (perhaps together with zone 3) plays a sensory and stimulatory role. The rich supply of large hairs, and the observation that the complementary zones of the female tremble throughout copulation, suggest this interpretation.

The midge Limnophyes pusillus (Chironomidae), like most midges, carries out the transfer of sperm from beginning to end in free flight. The contact elements of the genitalia are illustrated in Figs. 118 and 119. In principle, they are constructed like those of Tanytarsus discussed above:

Contact zone 1:

The end processes *ep* of the male hypopygium are flexibly jointed with the basal segments and normally are held in a folded position. During copulation they anchor themselves by pinching the eighth tergite of the female tightly between the rigid styli at their tips. The sides of the tergite are thereby pressed together and the middle is raised into an arch.

Contact zone 2:

The convex inner lobes *il* on the basal pincer segment *bp* of the male interlock very precisely and firmly with the correspondingly concave surfaces of the female lateral lobes.

Contact zone 3:

The anal projection of the male hypopygium is pushed into the female subgenital sinus.

Contact zone 4 (not illustrated):

The tip of the penis is laid on the unpaired opening of the spermatheca duct.

Here the exact relationship of the critical contacting elements, in zone 4, is set by contact zones 2 and 3, and the safety lock is provided by the clamp arrangement of zone 1. The contact surfaces 2 and 3 conform closely to one another and prevent displacement in either the longitudinal or the transverse directions; the position assumed is fixed by the pincers of zone 1. The cerci *ce*, on the other hand, apparently have no mechanical or sensory function in copulation; they lie between the inner basal margins of the basal pincer segments *bp*. The great density of hairs in zone 2 suggests that this zone may play an additional sensory role.

On the whole, the manner in which the connection is made in Limnophyes, which is completely adapted to copulation in the air, seems somewhat more fully automated than that of Tanytarsus, which may complete its copulation on the ground. No lever action is required for the parts to slide reliably into place (in Tanytarsus the medial and inner appendages must be actively positioned and pushed in); rather, the partners simply fit together an external complementary pair (zone 2), which in turn guarantees the proper positioning of the internal complementary pair (zone 3), as in the insertion of the electrical plug in Fig. 120. Active closure is thus replaced by a more passive seating mechanism, the critical element being forced to enter the proper channel when the guiding elements are in place. Once the connection is made the entire system, as in Tanytarsus, is locked together by the outer hypopygial pincers (zone 1).

The analogy with the firm connection of two electrical conductors by means of a plug and socket (120, 121) is more complete than one might suspect:

Category	Electrical plug and socket	Midge copulatory organs
Transmitting organs	Prongs and contacts	Penis and genital chamber (zone 4)
Internal complementary pair	Prong geometry and contact holes	Anal projection and subgenital sinus (zone 3)
External complementary pair	Guides *f* and grooves *n*	Inner lobe and lateral lobes (zone 2)
Other external mating parts	Geometry of plug housing and socket receptacle	Inner basal segment and outer surface of segment VII
Safety lock	Spring clip and socket catch *c*	Hypopygial pincers and geometry of segment VIII

The genital appendages of the various species within a genus show only slight differences, but these are sufficient to make impossible an exact fitting together of the genitalia and thus to prevent copulation between different species. Only the genitalia within a single species, which fit together in the minutest morphological detail, can complete the critical connection; this principle of specificity is identical to that of the lock and key.

In the yellow-fever mosquito Aëdes aegypti, the process of copulation has been investigated in detail. The male grasps the flying female from the back, attaches himself firmly with the hooks on his tarsal segments, swings down against the ventral side of the female and closes the contact between the genitalia (122–125). His claspers *cl* grasp the female cerci *ce* so that the tongue-like plate below the anus is lifted up and the vaginal opening exposed. With the anal hooks *ah* which lie opposite the claspers the male positions his genital opening in the ventral genital region of the female and seals the connection by muscular contraction. Very shortly thereafter the copulatory organ *co*, which has thus far been withdrawn, is protruded until the teeth at its tip mesh with the chitinous teeth on the dorsal margin of the vagina and pull it downward. Dilator plates at the end of the male copulatory organ thereupon widen the vagina and enlarge the opening of the bursa. Now the openings of the two organs are perfectly aligned and the alignment is mechanically reinforced. Immediately a quantity of seminal fluid containing about 2000 sperm is ejaculated into the bursa *bu* and, without delay, the above sequence of connections between the pair is released in reverse order; the two animals separate. The entire procedure is usually enacted during flight and lasts 14–20 seconds. In general, in all mosquito species which copulate in flight, the process of sperm transfer is begun by the fitting together of two matched parts, the geometry of the male (inner) part conforming exactly to that of the female (outer) part. The active positioning is done by the male.

From a biomechanical point of view, the mating of dragonflies and damselflies, quite a number of which also copulate in flight, proceeds in the reverse manner. The male copulatory organ does not lie in the usual place, near the orfice of the male gonads on the ninth abdominal sternite, but far anterior, on the ventral side of the second abdominal segment. It is a complex structure, evolved from the second and third sternites, and is sometimes segmented and shaped like a penis.

During courtship but before the actual copulation, the male empties sperm into this organ by bending the abdomen forward and bringing the genital opening of the ninth segment into contact with it. Also during courtship, the male flies ahead of the female, holding her firmly by the neck with clamp-like appendages of his eleventh abdominal segment. After the male has filled his copulatory organ with sperm, the female bends her abdomen forward and actively fits her external genitalia over the peculiar secondary male copulatory apparatus so as to take up the sperm. Near bodies of water in early summer one may frequently observe the flying wheel- or heart-shaped figures formed by pairs of copulating dragonflies.

This mating sequence involves three types of suitably matched structures: (1) male gonadal orfice—male secondary copulatory organ, (2) male abdominal appendages—female neck region, and (3) female external genitalia—male secondary copulatory organ.

Here again, the conformity of the male and female elements involved is strictly species specific. The male abdominal appendages are fastened either to the prothorax alone (damselflies) or to both head and prothorax (dragonflies) of the female; the teeth and other protuberances of these appendages fit like pegs precisely into indentations at the appropriate positions on the conspecific females. Such lock-and-key systems are so characteristic of each spicies that they are used by systematists to describe them. If an attempt is made at interspecific mating, it is usually given up even before copulation is tried.

Copulation itself lasts from a few seconds to 30 minutes. A damselfly female, for example, anchors her abdomen in position with the inner pair of gonapophyses, the ends of which are inserted far into the median invagination of the male copulatory apparatus, dorsal to the so-called hamuli anteriores of the male. The surfaces which catch and guide the gonapophyses are curved toward the ventral and lateral sides, so that the penis can enter without hindrance. Once again, there is an inner fitting of complementary parts and an outer locking device, and again even these are species specific—a second insurance against interspecific crossing. In the large dragonflies the manner of connection is more complicated still. There are at least two types of attachment devices. The combination has great tensile strength, and this is no doubt necessary, since the musculature of the female abdomen is certain to become fatigued during the long copulation of many dragonflies. Further details and illustrations of these remarkable adaptations are given by Pfau.

The dragonflies can release the copulatory coupling very quickly. In contrast to the damselflies, which usually form a "mating chain" and then land on the ground for copulation, many dragonfly species carry out the entire reproductive act in flight. Flying in tandem, the pairs are much less maneuverable and hence more vulnerable to their aerial enemies, the birds. Rapid "undocking" can be essential to survival. Here there is a nice correspondence between the specific design of an organ and the ecological circumstances of a species. The valves and other elements for the transfer of fuel to fighter planes are also designed so that when danger threatens the coupling can be broken with very little delay.

In summary, all of the connections in this category are based upon a single general principle: a rod-like component is introduced axially into a corresponding tube-like component. The connection must be secured against three possible kinds of displacement: (1) rotation about the longitudinal axis, (2) tilting with respect to the longitudinal axis, and (3) disengagement along the longitudinal axis. The physical arrangements which meet these requirements are varied and may involve the principles of (1) tongue-and-groove, (2) guide channels with matching surfaces, and (3) external clamping or internal anchoring.

2.1.2 Hook and Eye

One can also connect two parts by engaging a hook on one part with an eye on the other. This couples the parts securely with respect to tensile stress (point 3, above) but not with respect to tilt (2) and rotation (1). Tilt can be prevented by suitable guide surfaces on the two parts, and rotation eliminated by locating two hook connections of the same type next to one another with

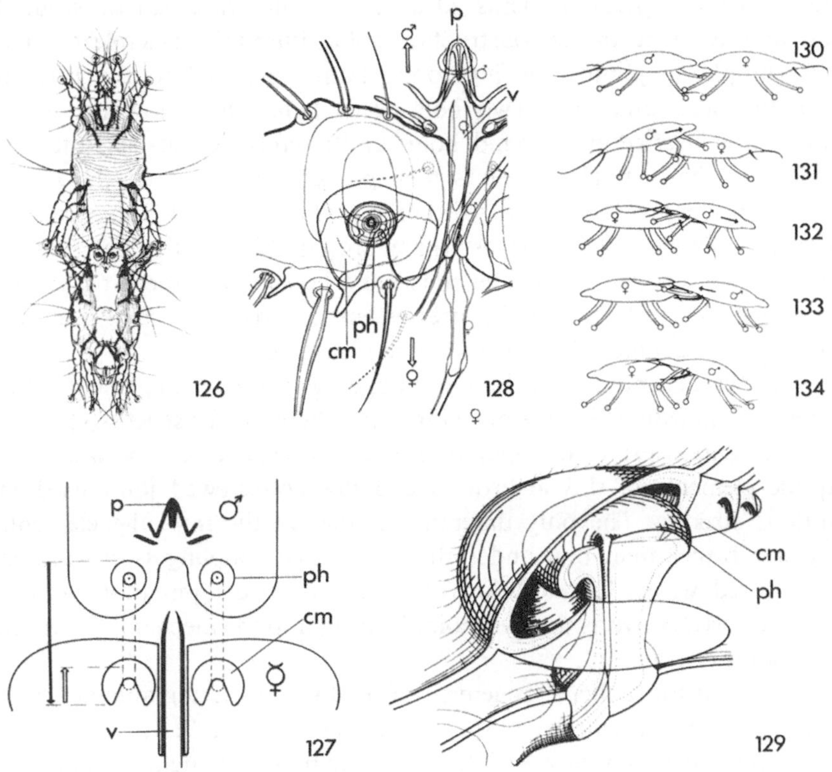

Plate 11 (Figs. 126–134)

126. Tritonymphs of Pterophagus strictus in copulation
127. Diagram of the connections involved in (126)
128. Semidiagrammatic drawing of the anchoring structures in Pterophagus strictus
129. Capstan/capsule elements in Pterophagus strictus
130–134. Adoption of the copulatory position in Pterodectes

the greatest possible separation. Thus, ideally, the connection can be secured against any displacement. In the construction of furniture this principle is often used—for example, in joining the horizontal frame of a bed to the vertical supports of the head and foot. Two bolt heads lying one below the other on the frame are inserted into round holes at the ends of slots let into the head and foot of the bed. The frame is then pushed down so that the heads of the bolts slide behind the flanges of the slot; the head and frame cannot be pulled apart. The parallel surfaces which prevent tilt are the flat mating surfaces of the two wooden parts. The slots must be suitably oriented; in this case they must be vertical, so that the stresses encountered in normal use tend not to release the attachment but to counteract disengagement.

Fig. 127 shows schematically a biological analog of this arrangement—the coupling mechanism with which the male bird mite Pterophagus strictus attaches himself to the female prior to copulation. The correspondence between the flange/capsule elements of this arthropod and the slot-screwed joint used in cabinetwork is striking. The only difference is that in the mite the elements lie side by side rather than one under the other. The coupling is secured in the three required ways. The addition of a central spacer pin (not shown), displaced posteriorly, converts the two-point bearing into the mechanically more stable three-point bearing.

Figs. 126–129 illustrate the arrangement of the two copulating animals, with a semidiagrammatic reconstruction of the connecting organs (129). It is necessary that the attachment between the copulating partners be strictly defined spatially. These mites live in the feathers of birds and are subjected to violent mechanical stresses when the bird flies; the delicate copulatory organs, the penis *p* and vagina *v*, are themselves not adequate to hold the animals together. Rather, they must be held together in a mechanically stable way, so that they cannot be shaken apart, before copulation can begin at all. The bird mites move slowly, dragging themselves along the barbs of the feathers, and their sensory capacity is very limited. The mechanism that locks them together is adapted to this way of life and automated accordingly. The male encounters the female head-to-head and pushes himself past, over the female's abdomen (130–134). The male has rounded, flanged protuberances *ph* like capstans on the posterior ventral surface, while the posterior end of the female has dorsally located capsules with a triangular section *cm* cut out. As the male's abdomen drags over that of the female, these paired elements are caused to slide together and lock in place. Only a few seconds of crawling are required for this connection to be completed. The broad, V-shaped cutout in each capsule of the female acts as a guide for the corresponding "capstan" of the male (128, 129), and the orientation of the slot is functionally correct, in the sense that the connection is reinforced automatically as long as the female continues to crawl forward and drag the male along. It can be released only by a brief backward movement of either partner.

2.1.3 Snap Fasteners

Snaps consist of a "male" part which is a rounded projection or peg expanded at the end (136) and a "female" part with a socket of the same diameter as the peg (135). The inner rim of the socket is reduced slightly in diameter by some spring arrangement. As the peg enters the socket it spreads the spring, and as soon as the broadest part of the peg has passed it closes again (with a "snap" or click), holding the peg firmly in the socket. Many modifications of the snap mechanism occur in engineering. The socket of a snap may be built directly into a component made of elastic synthetic materials, or it may even be replaced functionally by an opposed pair of leaf springs. An example from the animal kingdom is the margin of the squid mantle, which is joined to the body by means of two snap fasteners (137). Below the funnel there are two cartilaginous pits *b*, into each of which a sort of stud *g* protruding from the mantle wall can be pressed. This kind of snap corresponds functionally to the mechanism by which the legs of plastic dolls are held in their sockets. The opening of the mantle is thus sealed off after water has been drawn in through it. Now if the mantle musculature contracts vigorously, a high pressure is produced within the cavity, against which the snap fasteners must hold the mantle closed. Thus the pressure can only be released by outflow of water through the narrow funnel, which produces a considerable recoil, propelling the animal backward. Among the various families of squids the matching cartilaginous structures on funnel and mantle have characteristically different shapes (137–141). In some genera, for example Symplectoteuthis and Grimalditeuthis and in particular the members of the Cranchiidae, both of these snap fasteners have become firmly fused. In some Oegopsidae there is a second well-developed snap apparatus located in the neck region. It consists of a longitudinal rib of the dorsal rim of the mantle which is bent outward at right angles and can be seated in an elastic longitudinal cleft in the body (142, 143). The powerful collar muscle is attached to this neck apparatus; like a valve, it blocks the opening of the mantle so that expelled water must find its way through the funnel.

Squid of the genera Onychoteuthis, Abralia, and Galiteuthis have still another system of about half a dozen snap fasteners. The suction cups of the two long tentacles have been modified into claws. When the clawed tentacles grasp an animal, it is held firmly by this system of multiple hooks. But the connection can be released if the prey succeeds in pushing the tentacles apart. An ingenious system prevents this: at their bases the tentacles have a number of studs and sockets (144, 145) which are so arranged that when the two tentacles are laid together the male and female parts of each of these snap fasteners meet. In this way the tentacles are attached to one another at the base, and thus have a more peripheral fulcrum for the action of holding the prey.

Another snap apparatus which is bilaterally arranged is that with which some crabs (Decapoda, Brachyura) attach the abdomen to the ventral side of

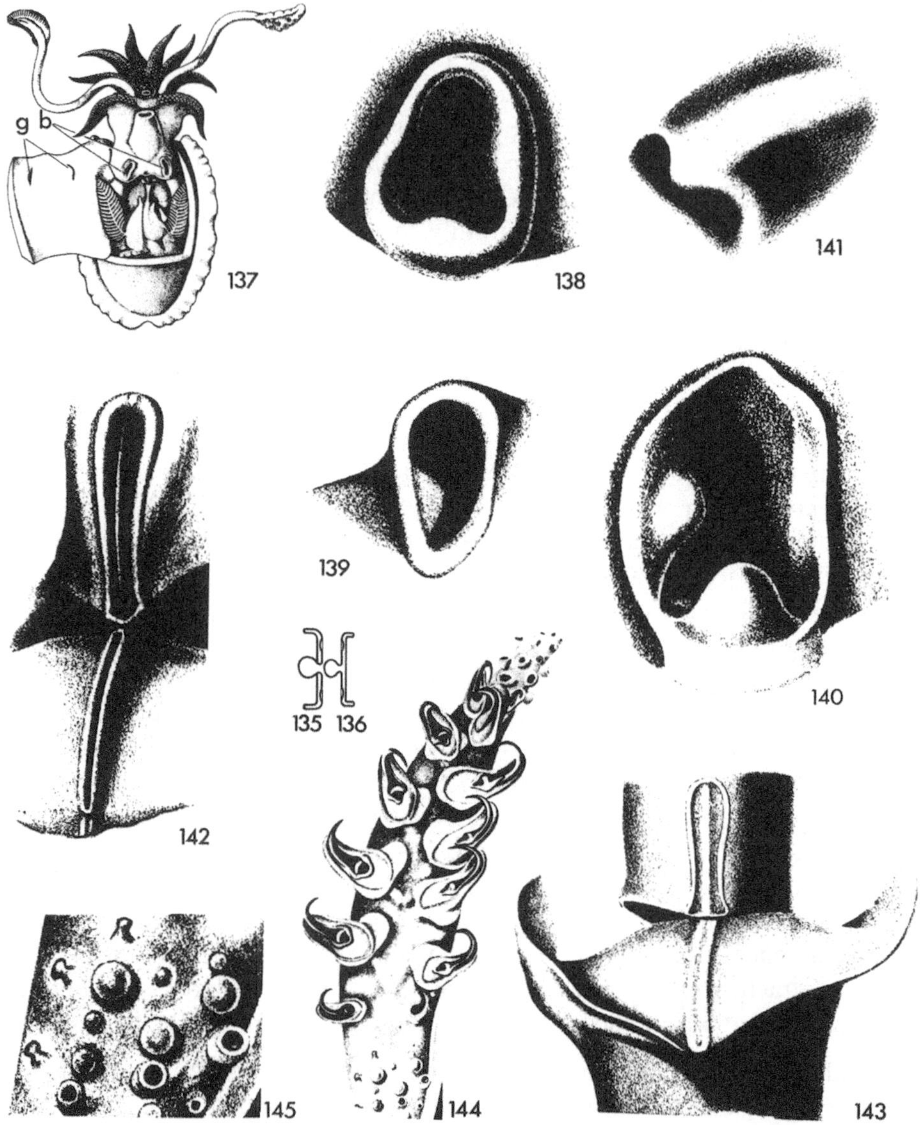

Plate 12 (Figs. 135–145)

135. Snap fastener, female part
136. Snap fastener, male part
137. Snap-type connection in Sepia officinalis
138 and 139. Cartilaginous pits on the funnel of Mastigoteuthis flammea
140. Cartilaginous pits on the funnel of Chiroteuthis imperator
141. Complementary part of the "snap" in (140)
142. Elongated snap fastening in Benthoteuthis megalops
143. Chiroteuthis imperator
144. Tentacular fasteners
145. Enlarged section of (144)

the thorax. They carry the reduced abdomen bent forward and pressed into a shallow depression in the thorax. The fifth thoracic sternum bears a papilla on either side of the median line, which acts as the male part of the snap. The corresponding female parts take the form of appropriately arranged sockets on the sixth abdominal sternum. This device maintains the desired position without the necessity for continuous contraction of the tail muscles.

An exceptionally curious reciprocal snap fastener is found in trematodes of the genus Diplozoon. These parasitic flukes, up to 10 mm in length, live on the gills of fresh-water fishes. They exist as so-called "double individuals"; the adult animal results from the fusion of two larvae. Each single larva develops a ventral sucker (primordium of the cirrus sac) and a dorsal papilla (the opening of Laurer's canal). Two larvae lie together as shown in Fig. 146; the ventral sides are indicated by the hatched and dotted shading and the dorsal sides are unshaded. In this position they make a connection between papilla and sucker. Then the posterior parts of the two bodies are crossed over and each is twisted 180° so that the second snap connection can be made (147). Thus united, the animals undergo secondary modifications (148) and finally attach themselves to a row of gill lamellae in the host fish, never to separate again. The two hermaphroditic animals can now exchange their male gametes undisturbed, since the vas deferens of each "half animal", which opens into the cirrus sac and out of which the sperm threads emerge, is securely snapped into the Laurer's canal of the other, which opens into the oviduct near the ovary. Thus a direct passage of sperm to the eggs, still without shells at that point, is guaranteed (149).

From another point of view, this double fastening of two individuals has been considered as an adaptation to life in the current of water which flows between the gills. The advantage arises because the two hemispherically shaped posterior ends (150) can be held together (like a hinged double spoon for making balls of ice cream) so as to enclose a certain volume of the gill material. The double individual is therefore much less likely to be washed away than is a single animal. In this sense, the double-snap attachment would correspond to the pivot pin which holds together the two halves of a pair of pliers. In favor of this interpretation is the fact that double individuals are regularly found on areas of the gills washed by turbulent water, while single individuals manage to live only in much less exposed regions of the gill.

A very well developed and effective snap mechanism locks the hemielytra of the water bug Nepa firmly onto the body. Again, the fastening is a double one. On the epimeron of the mesothorax there is a button-shaped thickening which fits securely into a corresponding cavity on the anterior margin of the hemielytron (151, 152). In addition, the side of the abdomen has a long, sharp marginal ridge which fits into the grooved margin of the hemielytron. Thus the air space beneath the wings, which is essential for underwater respiration, is effectively sealed. More or less well-developed elytral seating mechanisms

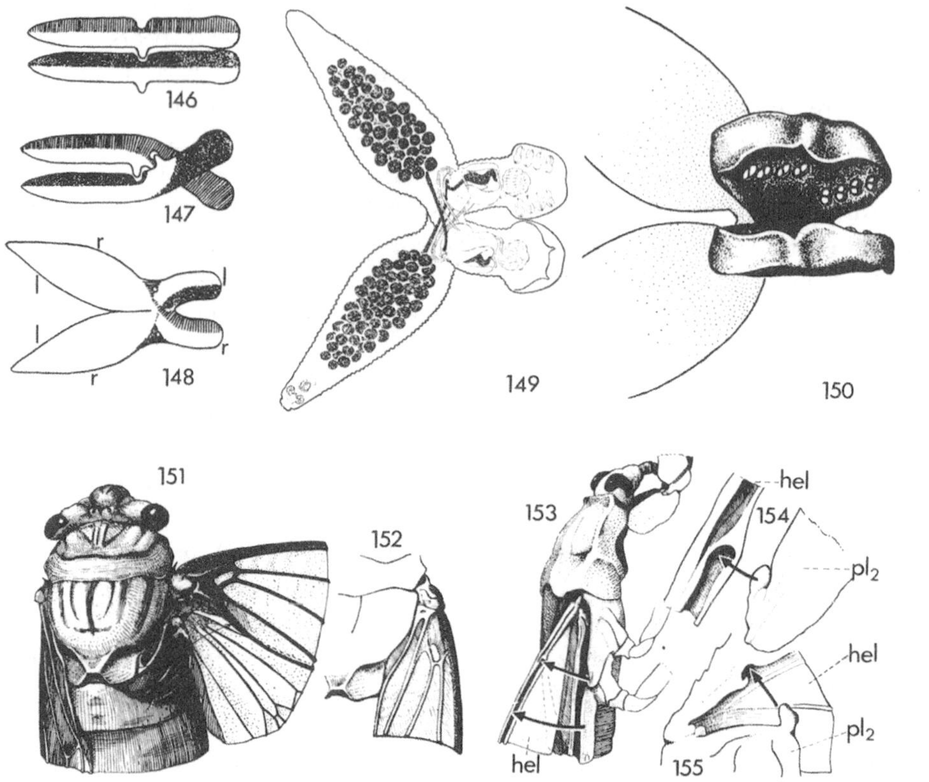

Plate 13 (Figs. 146–155)

146–148. Formation of a fused pair in Diplozoon. r, right side of the body; l, left side

149. Details of the "double individual" Diplozoon

150. Diplozoon on the gill lamellae of a fish

151 and 152. Snap-type fastening in Nepa

153. Hemielytral locking device in Gelastocoris

154 and 155. Snap-type fastening in Queseda gigas. hel, hemielytron; pl_2, pleuron of the second segment

of this sort are found in almost every water bug, including the well-known back swimmer Notonecta. In the "water boatman" Corixa and in the genus Gelastocoris, the seating mechanism operates in the reverse way—the hemielytra *hel* bear the female elements and the metepimera, the male elements. But the sockets are open laterally, so that the protrusion is more "hooked in" than "pressed in", and when seated is locked in place (153).

Cicadas also have evolved astonishingly powerful snap mechanisms (154, 155). In the South American species Queseda gigas, according to Weber, "the strongly sclerotized, proximal posterior margin of the forewing, which projects ventrally like a ridge, snaps so firmly into a corresponding groove in the lateral part of the metascutellum that a certain amount of effort is required to move the wings of a dead cicada into flight position." The biological significance of this mechanism, which also fastens the forewings to the scutellum in the Cercopidae and most terrestrial bugs, is not really clear.

2.1.4 Multiple Connectors

A zipper consists of two rows of metal or plastic elements laid in parallel, but with the rows displaced so that the elements are staggered. In one design, each of these tooth-like elements has a stud on one side and a socket on the other. Each element can thus fit into two adjacent elements of the opposite row. An appropriately designed slide forces them to interlock or to separate. Another type of zipper operates according to an "interlocking chain" principle. Perlon spirals are sewn onto the free edges of the pieces of material to be joined so that the free side of each spiral forms a row of parallel loops. These, like the rows of hooks in the first type, are engaged and disengaged by a central slide. Figs. 156 and 157 show these two basic types. The common principle is that identical interlocking elements are arranged in series. This permits sheets of material, usually fabric or foil, to be joined securely along their edges and to be separated rapidly again.

In the animal kingdom there is no case in which two elements are joined and separated by means of a third sliding element, but connecting devices similar to zippers do occur. For example, each element may have half a dozen teeth, positioned so that the teeth in one row oppose the gaps in the other; functionally, the situation is intermediate between the two types of zipper described. Fig. 158 shows a beetle mite (Oribatei, Acarina). The sides of these animals are wedge-shaped. The two solid, central longitudinal plates on the ventral surface, the so-called anogenital plates (159), are coupled in the middle by a releasable multiple connection. This attachment device is shown enlarged for the species Euphthiracarus cribarius (160) and Rhysotritia duplicata (161); in both cases the "zipper" is unzipped. When closed, the surface of the interlocked elements has a braided appearance (159).

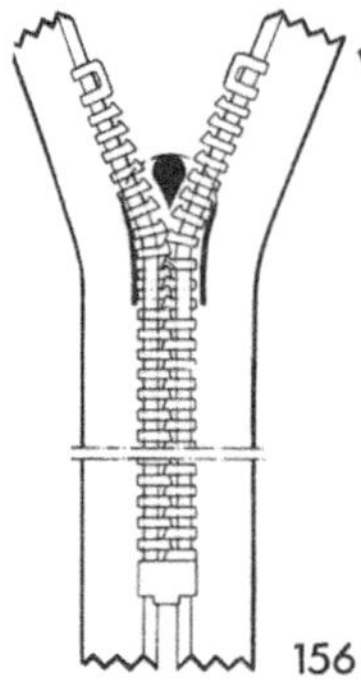
156

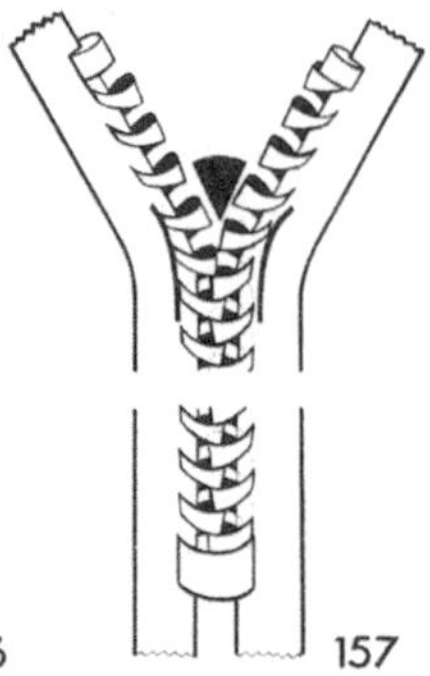
157

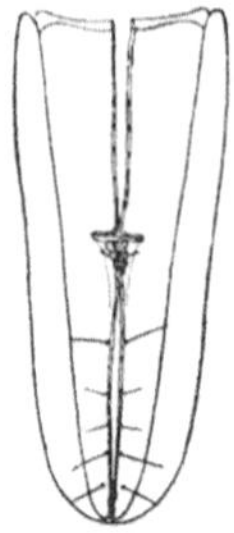
159

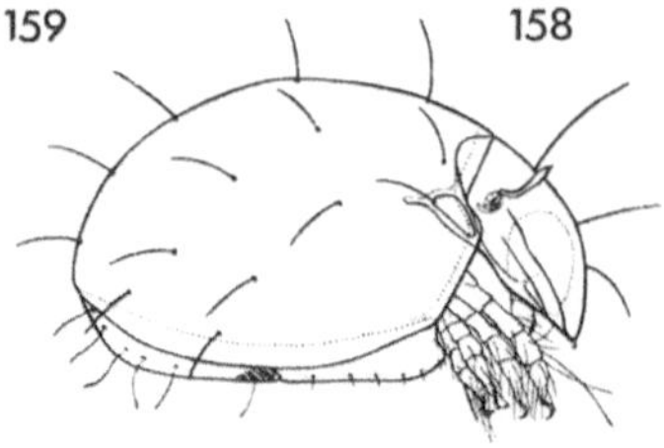
158

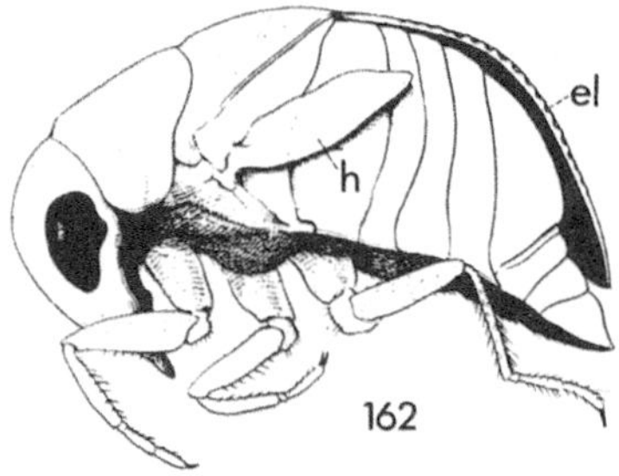

162

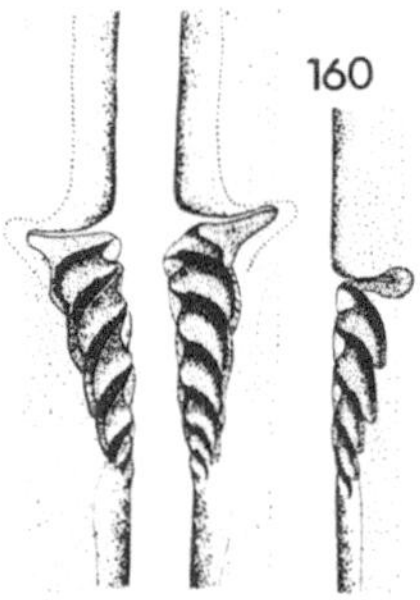
160

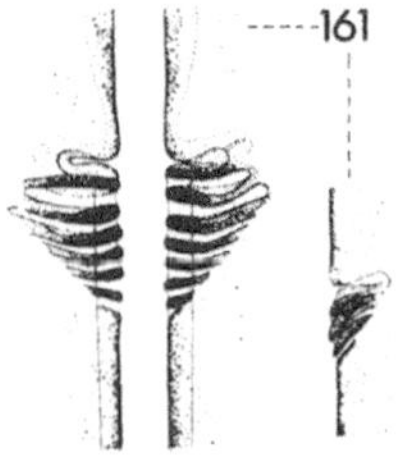
161

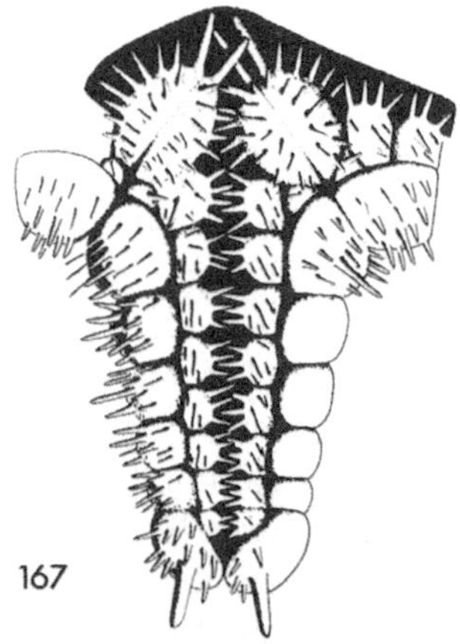
167

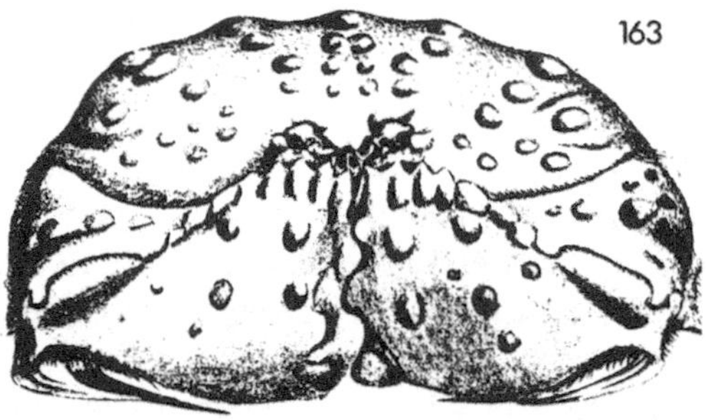
163

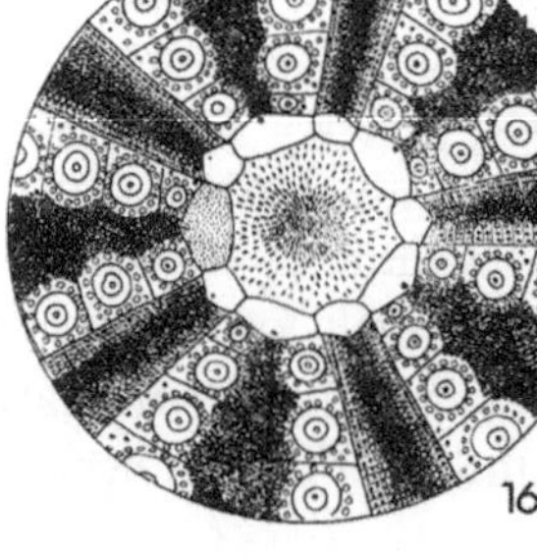
164

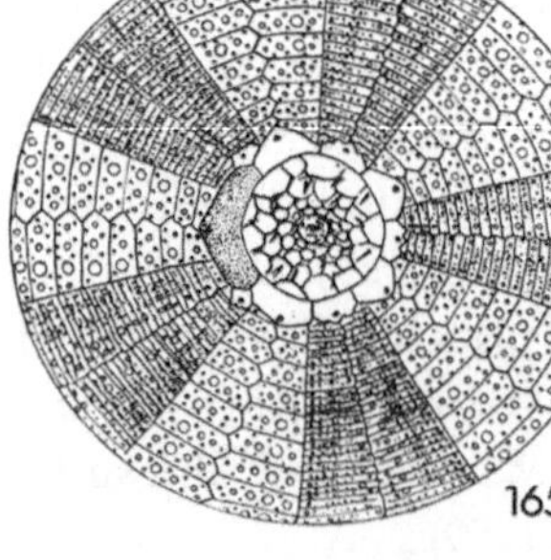
165

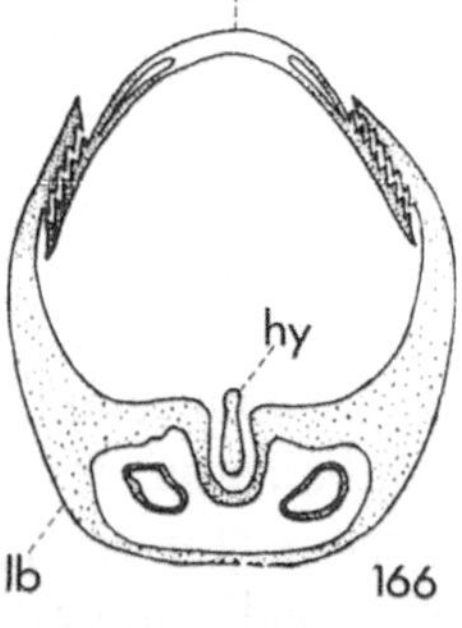

166

There is an excellent multiple connection between the two hemielytra of the North American water bug Plea striola. The elytra are very long and encase almost the whole abdomen. They meet at the midline; over the entire region where the two edges touch they bear appropriately staggered rows of teeth (162) which fit into one another exactly. Approximately 20 zipper-like elements suffice to lock the elytra firmly together.

In a similar manner, though in this case with only two or three teeth, the serrated "hands", or propodites, of both claws of the mediterranean crab Calappa granulata can be wedged under the carapace. With the claws in this resting position, the animal becomes a completely enclosed armored hemisphere (163). The connection can be released without effort at any time. Such multiple interlocks between triangular elements also join the segments of certain rigid shells. This is true, for example, of the fragile ambulacral plates, and in particular of the strong interambulacral plates, of the sea urchin (164, 165). This arrangement is part of the mechanical design which accounts for the remarkable compressive strength of the sea-urchin test.

A different principle is applied in the connection between labium and epipharynx in the piercing proboscis of the stable fly. On each side, there are about half a dozen marginal ridges which interlock. In cross section these zippers have a sawtooth appearance (166). The females of the Australian sawfly (the symphytan hymenopteran Perga affinis) have a five-part egg-laying apparatus at the end of the abdomen. On its edges are rows of sawteeth used for sawing a hole in which to lay the eggs; at rest the teeth presumably interlock alternately.

The ventral ambulacral plates of the arctic deep-sea starfish Cheiraster gerlachei interlock by means of three powerful spurs on each plate which are directed towards those of the adjacent plate (167). This junction must have the effect of counteracting any relative longitudinal displacement of the two halves of the arms, but it is not known whether this is the actual significance of the device.

The two halves of the siliceous skeletons of certain Radiolaria are wedged together by a peculiar multiple "plug" mechanism. This is basically a multiple arrangement of hooks and eyes, in which spines on one side fit into pits, holes,

◀ Plate 14 (Figs. 156–167)

156. Toothed zipper
157. Spiral zipper
158. Anogential plates in Eupthiracarus monodactylus
159. Anogenital plates in Rhysotritia
160. Anogenital plates of Eupthiracarus cribarius, ventral and lateral view
161. Anogential plates of Rhysotritia duplicata, ventral and lateral view
162. Elytral fastening mechanism of Plea striola.
163. Toothed interlock of propodite and carapace in Calappa granulata
164. Eucidaris tribuloides
165. Tripneustes ventricosus
166. Fastening of labium (lb) to epipharynx (e) in Glossina. hy, hypopharynx
167. Interlocking of ambulacral plates in Cheiraster gerlachei

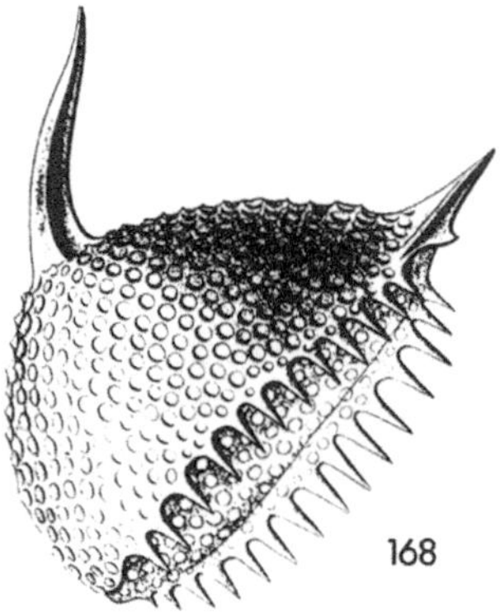
168

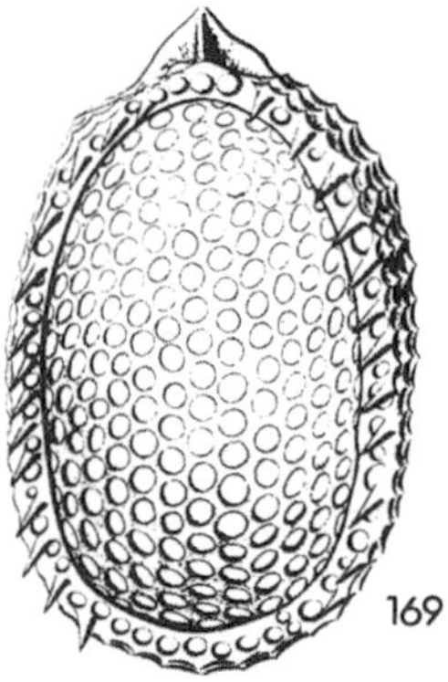
169

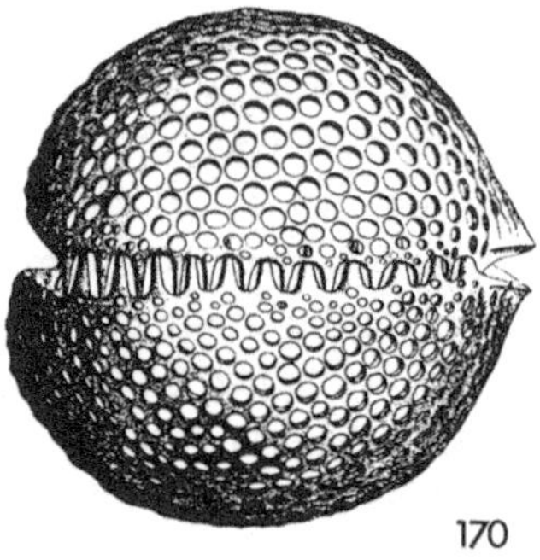
170

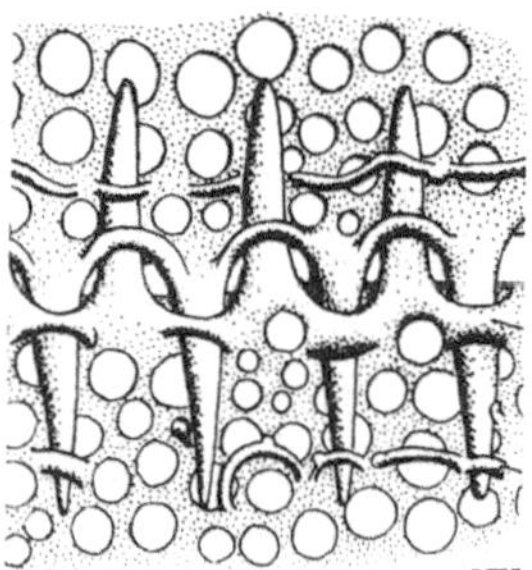
171

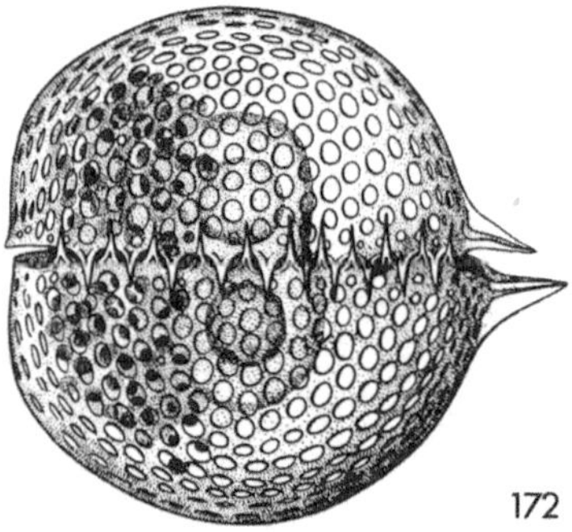
172

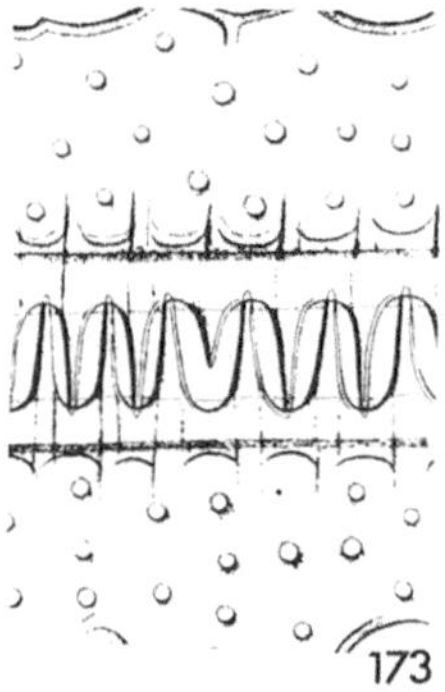
173

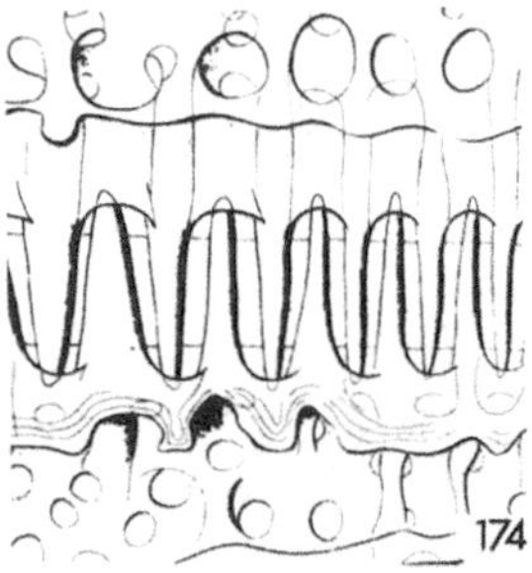
174

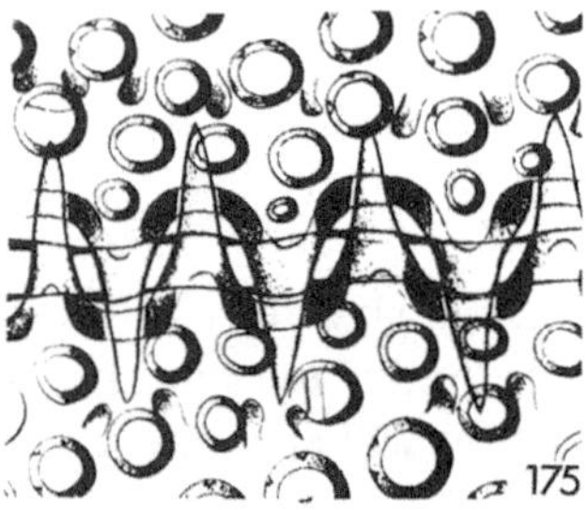
175

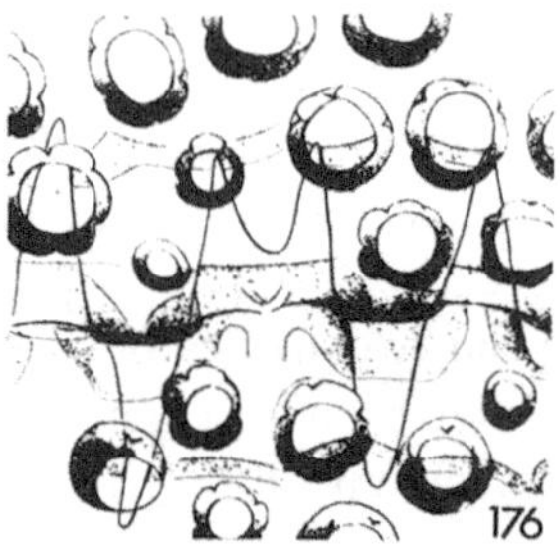
176

177

178

rings or lugs on the other. On each side, hooks and eyes alternate with perfect regularity. In correspondence with its position at the periphery of a hemisphere (168), each such system is arranged in the form of a part of a circle or ellipse (169). Some examples are shown in Figures 168 to 176. These two-part spherical lattices joined by interlocking rows of teeth are found in the benthic tripylean Radiolaria (suborder Phaeoconchia, family Concharidae). In Conchidium (168) and Conchoceras (171) the teeth extend inside the opposite shell half, which bears on its inner side two arches for each tooth—a sturdy one for the base of the tooth and a fine one for its tip. This delicate row of fine arches can also be modified into pairs of pillars which do not grip the tip of the tooth but simply guide it, one on each side. In either case any displacement tangential to the rim of the hemisphere is absolutely prevented. On the other hand, there is allowance for the two halves to separate slightly without disengaging as the soft body of the animal grows.

Other multiple connections in Radiolaria are more simply constructed. The marginal spines of Conchidium argiope extend not within, but upon, the opposite shell, like the fingers of one hand when they are intermeshed with those of the other (172). Sometimes a sort of ridge or pocket is provided as an abutment for the tip of the tooth, as in Conchopsis (173, 174), and sometimes the margins of the shells at the base of the teeth are scalloped so that they fit tightly together upon closure, as in Conchidium tridacna (175, 176). The species name of the latter animal reflects the similarity of this scalloping to that in the shell of the giant (up to one meter long) clam Tridacna gigas and its smaller relatives (177). The interlocking shell margins of these clams can also close without a gap.

The surfaces of the cells in the stratum spinosum of human skin bear submicroscopic papillae, rather like rounded pegs. Their diameter is greatest (a few tenths of a micron) at the distal end, so that they have a capstan-like appearance. During cell development they are wedged together so that they form a two-dimensional system of multiply interlocking studs (178). There are technical equivalents of such systems, such as the nubbly strips of rubber which when pressed together resist tangential stress.

◀ Plate 15 (Figs. 168–178)

168. Conchidium triodon
169. Conchidium argiope
170. Conchidium terebrantula
171. Interlocking mechanism in Conchoceras caudatum
172. Conchidium argiope
173. Shell interlock in Conchopsis orbicularis, from outside
174. Shell interlock in Conchopsis orbicularis, from inside
175. Shell interlock in Conchidium tridacna, from outside
176. Shell interlock in Conchidium tridacna, from inside
177. Tridacna
178. Human stratum spinosum

Finally, a zipper-like connection is found on the ventral margins of the two semicylinders which form the proboscis of the hawk moth. It consists of a large number of double elements, arrayed in series and wedged together in pairs. A more detailed description is given on page 18.

2.2 Attachment by One Specialized Device

The attachment devices discussed in Section 2.1 required a precise alignment of the objects to be joined. In many cases, however, the purpose of the attachment is served even though the objects are not uniquely oriented. Here it is convenient if only one special connecting element is used, which can attach to any of a number of different places on the object to be held. The surface of the object, of course, must meet certain general requirements, which are different for the different types of connecting devices.

As an introduction to the kinds of couplings to be described in this section, the table below compares typical examples. Although most such attachments, as we shall see, involve several physical principles, the table at least suggests a framework for discussion.

Type	Man-made example	Biological analog	Surface requirement
1. Clamps	Vise	Prehensile legs of insects	Arbitrary
2. Grippers	Pliers	Clasping antennae of Anostraca	Arbitrary
3. Hooks	Anchor	Thorns of climbing plants	Irregular
4. Probabilistic fasteners	"Velcro" tape	Plant burrs	Irregular
5. Expansion fasteners	Wall plug and screw	Clam foot	Penetrable
6. Suction devices	Suction cup	Octopus sucker	Smooth
7. Gluing devices	Masking tape	Cement glands	Adequate adhesion
8. Intermolecular adhesion	Wet paper on glass	Tree frog pads	Adequate adhesion

It is convenient to treat these eight forms of coupling in the above order.

2.2.1 Clamps

One speaks of "clamping" when an object is held firm by a special device which exerts strong local pressure at opposed points, or a lesser pressure applied over opposed surfaces. An example of the first case is a vise, the jaws of which grasp an object with great force over a relatively small area, so that the local pressure is high. The shape of the object is unimportant. Devices of the second type include the split-sleeve clamp (188), the split collet of the toolmaker's lathe, and others, all designed so that the jaws conform closely to the round object to be held. Forceps, the third subclass of clamps to be discussed, are a special case of the first type.

2.2.1.1 Vises

The vise principle is embodied in the oddly constructed anterior tarsi of the female Dryinidae (Hymenoptera) (179, 180). All the tarsal segments are elongated. The fifth segment is particularly long; it extends out to the side and bears rows of projections ranging from tooth-like to lamelliform in shape. At the base of this extension a long hook is attached by a joint so that it can be folded against the fifth segment like the jaws of a clamp. Enclosed objects are pressed against the rough toothed ridge and held as though in a hand vise; as with most vises, the jaws can constrict or indent the object held so as to improve the grip. The monogenean trematodes (Platyhelminthes) are ectoparasitic flukes with posterior attachment organs of characteristic shapes ("opisthaptors") which anchor them securely to the gills of fish. It is characteristic of the diclidophoroid genera (for example, Microcotyle, Dactylocotyle, and Diplozoon) that their opisthaptors include an apparatus which operates like a paper clip; cuticularized projections are used to clip together several adjacent gill lamellae *g* of a host fish (182–184).

Figs. 185–187 show the forelegs of various ambush bugs (Phymatidae). The species Carcinocoris binghami folds the tibia *ti*, with its attached non-functional tarsus *ta*, against the femur *fe*, which is honed like a knife; this crab-claw-like prehensile apparatus is unique among insects. But other predatory bugs, particularly the Gelastocoridae, also possess effective prehensile legs.

2.2.1.2 Split-sleeve Clamps

The principle of the split-sleeve clamp (188) is used by many dipteran larvae of the family Ephydridae (189) to attach themselves to the stems of aquatic plants. Some genera of these families have a round indentation on the ventral side, near the posterior end, which has barrier hairs along the edge, directed inward; the clamp can be closed by means of a special muscle (190). Before pupation the larva siezes in it a cylindrical stem of a plant, usually of exactly

179

180

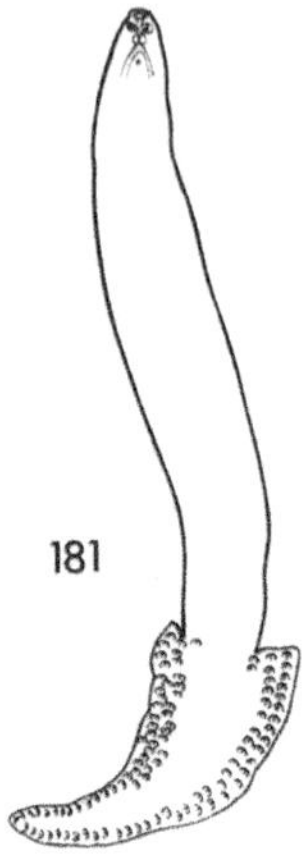
181

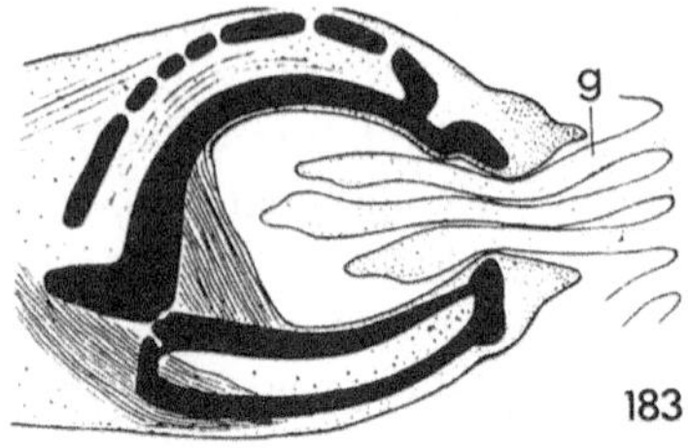

183

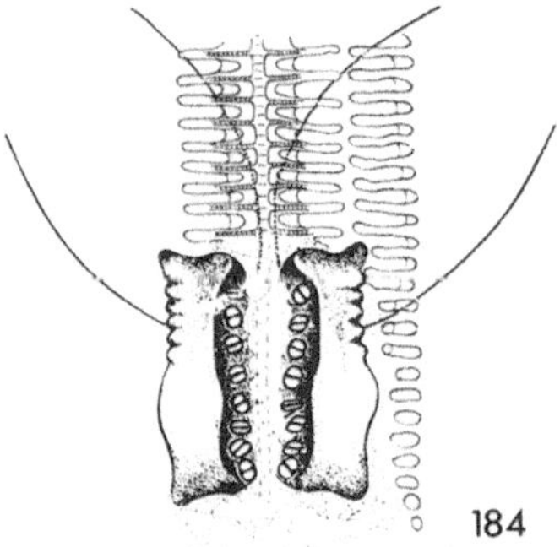
184

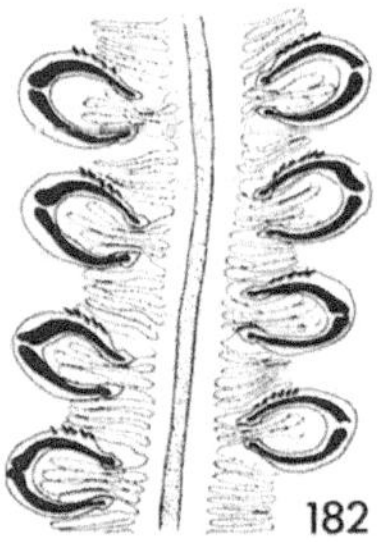
182

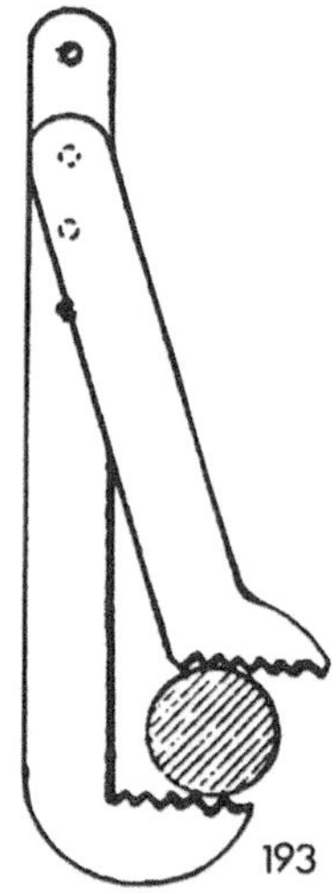
193

189

185

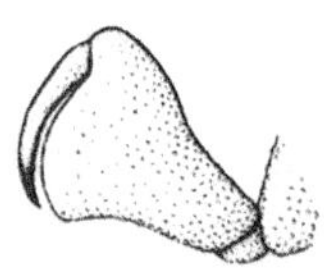
186

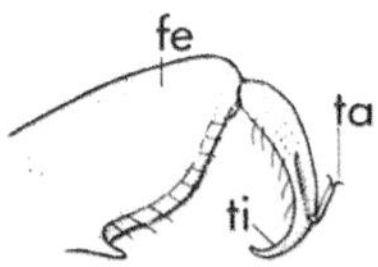

187

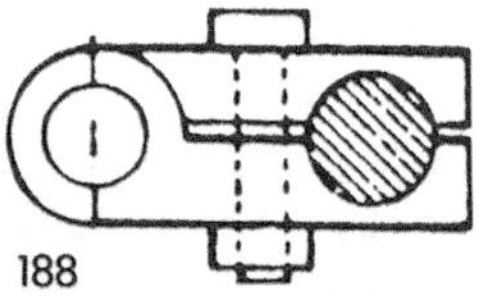
188

190

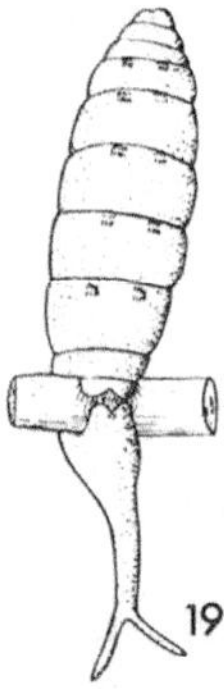
191

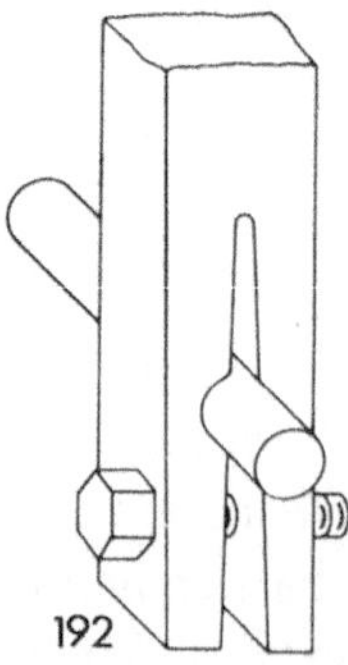
192

194

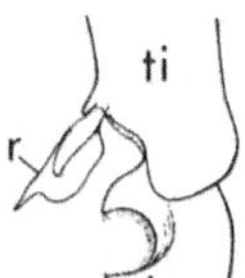

195

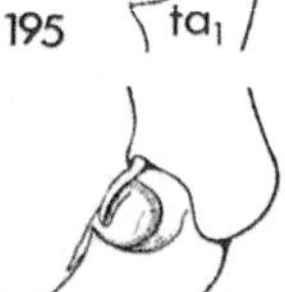
196

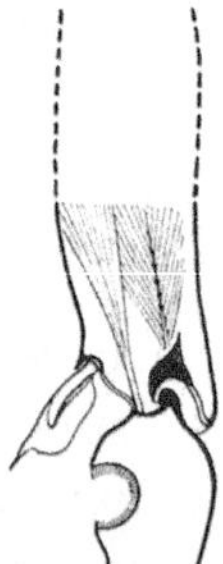
197

the appropriate diameter (191). Screw-clamps (192) and toothed pipe-wrenches (193) serve related functions.

The foreleg of the honeybee Apis mellifera bears an apparatus for cleaning the antennae (194). It consists (195) of a semicylindrical notch, of the diameter of the antenna, on the inner edge of the first tarsal segment ta_1 lined with a single comb-like row of close-set hairs, and a slightly hollowed-out projection of the tibia *ti*, the fibula *r*. When it is to be cleaned, the flagellum of the antenna is laid in the notch and the tibiotarsal joint is closed by muscles (196, 197). The antenna is now gripped firmly in the notch. When it is pulled through, the hair comb cleans the sensory outer surface and the fibula scrapes off the inner surface. A cleaning apparatus of this type has been developed in most hymenopterans, but in those the fit between it and the antenna is not as perfect as in the honeybee.

It is interesting that numerous species of carabid beetles, those belonging to the group of small, agile Harpalinae, have developed a quite similar antennal cleaning device. In these animals it takes the form of a hair-lined notch on the tibia. The antenna is guided by two sturdy spurs and is sometimes clamped into the notch by two long, springy hairs. The cleaning devices of the Hymenoptera and Harpalinae are a fine example of the evolution of analogous structures in the animal kingdom. Chirodiscus amplexans (198) uses semicircular modifications of the ends of the tarsi, which function quite similarly, to cling to the feather shaft of the host animal; the dimensions of the tarsi (199) are well adapted to those of the shaft. The device shown in Fig. 200, used to clamp rods in place in a drill press, is functionally similar to these biological mechanisms.

In some tetragnath spiders—for example, the species Pachygnatha clercki—the males immobilize the powerful chelicerae of the females prior to copulation. They grasp them with their own chelicerae in such a way that the outer segments of the female chelicerae fit into semicylindrical grooves, surrounded by spur-like processes, on the inner cheliceral segments of the male; the female's chelicerae are held in this position by the male's outer cheliceral segments, which fold shut like a pocket knife (201).

◀ Plate 16 (Figs. 179–197)

179. Anterior tarsus of a female dryinid wasp
180. Anterior tarsus of Lonchodryinus tricolor
181. Microcotyle pogoniae
182. Attachment clips of Diclidophora denticulata
183. Sagittal section of one clip (from 182)
184. Hind end of Diplozoon tetragonopterini
185. Foreleg of Phymata crassipes
186. Foreleg of Macrocephalus leucographus
187. Foreleg of Carcinocoris binghami
188. Split-sleeve clamp
189. Ephydrid larva
190. Larva of Ephydra subopacta
191. Puparium of Ephydra riparia
192. Diagram of a dowel clamp
193. Pipe wrench
194. Toilet apparatus of Apis mellifera
195. Toilet apparatus of Apis mellifera, opened
196. Toilet apparatus of Apis mellifera, closed
197. Closer and opener muscles for (194–196)

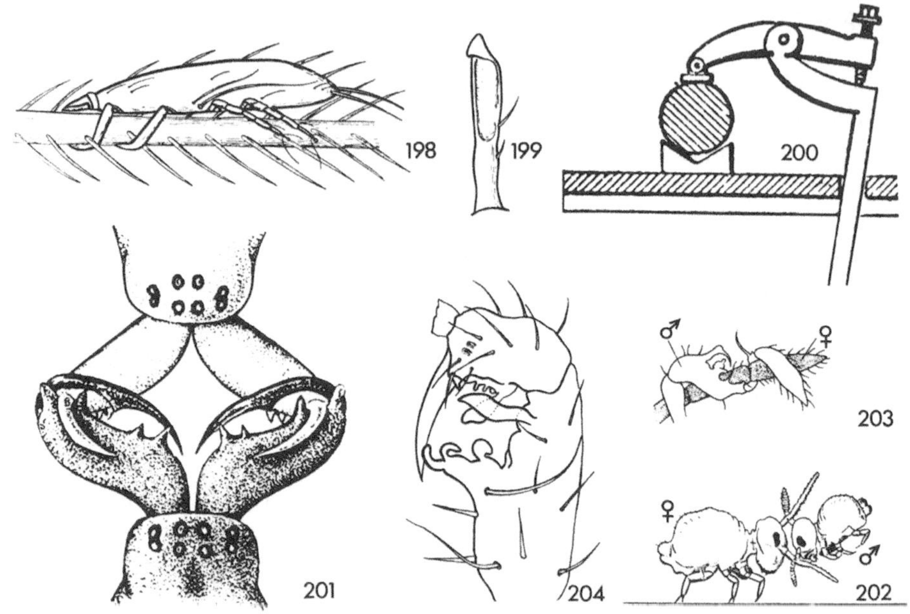

Plate 17 (Figs. 198–204)

198. Chirodiscus amplexans (Cheyletidae) on a feather of Podargus strigoides
199. Tip of foreleg of Chirodiscus amplexans
200. Mechanism for clamping a round object
201. Immobilization of the chelicerae in Pachygnatha clercki
202. Mating pair of Sminthurides aquaticus
203. Clasping antenna of Sminthurus apicalis
204. Terminal segments of a clasping antenna

Male Collembola of the genus Sminthurides attach themselves to the antennae of the females prior to copulation by means of specially constructed clasping antennae (202); sometimes the males are carried about in this way for days. This connection represents a combination of the split-sleeve and coiling principles (203). Numerous soft-skinned papillae and toothlike projections on the male antennae (204) fill in any irregularities of the mating surface and ensure a firm seating.

2.2.1.3 Forceps

There is a further large group of grasping organs and mechanisms, in both biology and technology, which is generally characterized by a fundamental forceps-like construction, modified according to the substrate which is to be grasped. For example, for soft material they might be shaped like broad spoons. Their form is often ingeniously adapted to the function. Often the object to be grasped can be seized and gripped securely even when hidden in a tubular flower (the humming-bird beak) or obscured by a bone (certain surgical instruments).

The two-jawed pedicellariae of the Forcipulata (sea stars, Echinodermata) are located singly or in bundles upon movable stalks. They often have delicate calcareous skeletons and serve primarily for the capture of prey—grasping and holding small crustacea with their finely toothed edges. The skeletal valves within the jaws of a pedicellaria are seated on a common basal piece at the end of the stalk. Unstalked pedicellariae occur in other orders of sea stars, the Phanerozonia and some of the Spinulosa. These are connected by a movable joint to ossicles of the endoskeleton. One type (205) consists of two opposed spatulate valves which lie spread out in basal depressions when open; when closed the teeth at the margins of the expanded, spoon-like jaws interlock precisely (206). A similarly good fit is achieved by the two-jawed prehensile mask of dragonfly larvae (207). The teeth of rodents are also particularly analogous to forceps and pincers.

The functional shapes of the beaks of birds often find striking parallels among the surgical instruments. The beak of the African openbill Anastomus lamelligerus (208) operates on the principle of an artery forceps (209), while the snipe head (210) is like the Grünewald-type conchotome (211), an instrument for removing the turbinate bones of the nose. When closed, its thickened tip acts as a probe (212). The bill of the northern diving bird Fratercula arctica, which is adapted for breaking up and crushing small but hard marine animals, is like Langenbeck's hemorrhoid forceps. The bent bill of the flamingo also has its analogs among the surgical instruments, and that of the black skimmer (213) is like a pair of angled cannula scissors (214). Compare the eagle's beak (215) with a pair of rib scissors (216), and the tearing beak of the vulture (217) with a rachiotome (218). These have similar functions and almost identical shapes; the two instruments are especially designed for operations on the vertebral column.

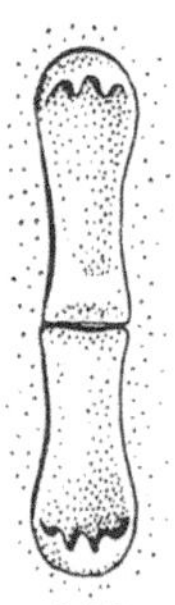
205

206

207

208

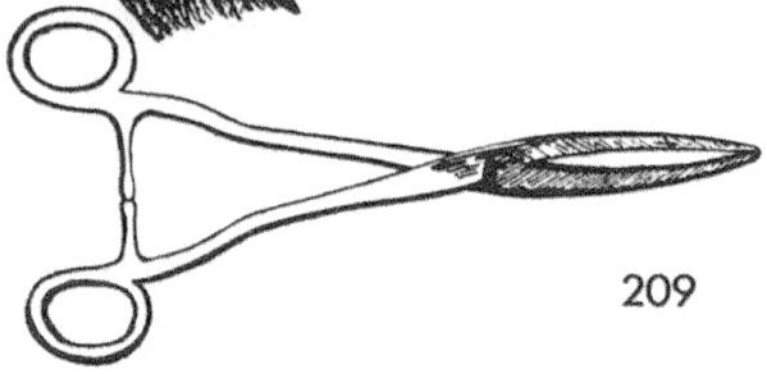
209

213

212

210

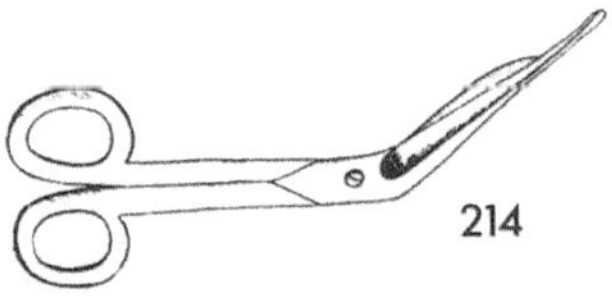
214

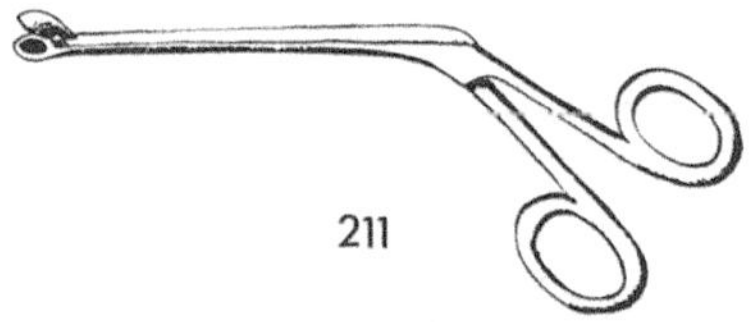
211

215

217

216

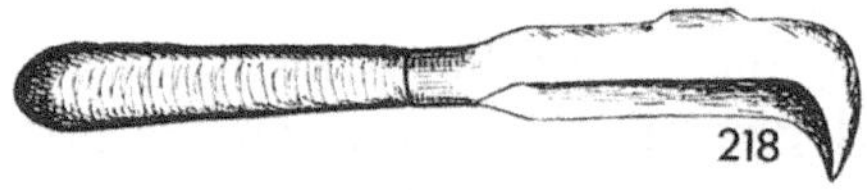
218

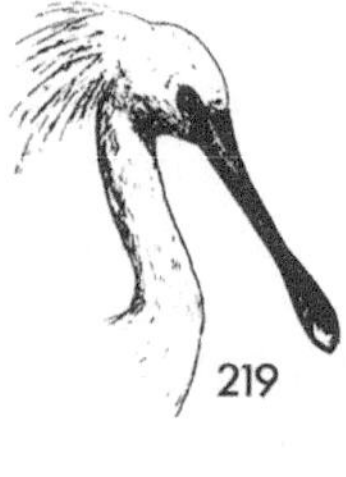
219

220

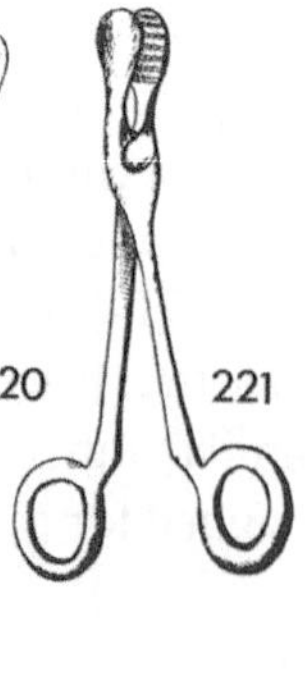
221

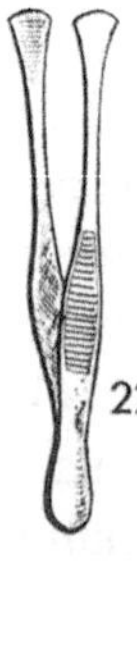
222

223

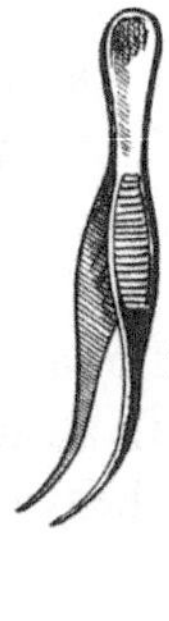
224

A number of other forceps-like instruments find biological analogs in the bills of birds. That of the spoonbill (219, 220) is similar both to a surgical forceps with broad flattened tips (221) and to the Danish type of cover-slip forceps (222). The sideways-curved beak of the New Zealand wrybill Amarhynchus frontalis (223) corresponds to the sharp, curved forceps for the removal of splinters (224); when closed the beak functions like a curved dissection needle.

The enormous variety among surgical instruments of course results from the great differences among the tasks for which they are designed; similarly, the horny beaks of birds correspond in structure to the extremely varied functions which they perform. Pulverizing, chopping, crushing, chewing, and tearing demand different structures than scooping, straining, sifting, lifting, and probing. Still other forms are required for chiselling, hacking, picking, stabbing, weaving, and sucking nectar. The bill shape technically best suited to a particular purpose has evolved through natural selection. The extraordinarily large number of surgical instruments, too, has arisen through selection of technological variations upon a few fundamental forms; modifications were tried out until an ideal form was found for each purpose.

Among the beetles there is a whole series of species which possess clamp mechanisms formed by the forceps-like interaction of the mouthparts, thoracic projections, or the forelegs. The palps of the mouthparts of the Japanese carabid Damaster blaptoides viridipennis have broadened, spoon-shaped ends which together form a four-jawed clamp (225); the male dung beetle Phanaeus imperator (226) from South America bears on its head a long spur, bent backward, which can be folded against two processes of the scutellum; the Spanish beetle Copris hispanicus (227) possesses a similar device. In the African dung beetle Heliocopris gigas, this clamp mechanism is formed in just the opposite way (228). Two lateral processes on the head, directed backward, can be pushed back against a medial thoracic process which is bent forward.

The male Brazilian rhinoceros beetle Enema pan (229) has a similar arrangement, but here there is only one thoracic horn curving forward. In the Hercules

◀ Plate 18 (Figs. 205–224)

205. Pedicellariae of a sea star (Spinulosa), open
206. Pedicellariae of a sea star (Spinulosa), closed
207. Mask of a dragonfly larva (Cordulegastridae)
208. Beak of Anastomus lamelligerus
209. Artery forceps
210. Head of a snipe
211. Conchotome
212. Probe
213. Skimmer
214. Cannula scissors
215. Beak of an eagle
216. Rib scissors
217. Beak of a vulture
218. Rachiotome
219. Beak of a spoonbill
220. Beak of a spoonbill, opened
221. Broad-tipped clamp
222. Cover-slip forceps
223. Beak of Anarhynchus frontalis
224. Splinter forceps

beetle Dynastes hercules (230) from Central America, which can be up to 17 cm long and also belongs to the family of the rhinoceros beetles, the thoracic process is extremely long and aims forward. In Dynastes neptunus (231), from Columbia, the long process of the pronotum is bent downward and covered thickly with hairs; the somewhat shorter cephalic process is bent upward and bears thorny protuberances. In Golofa portere (232), a Columbian rhinoceros beetle, the apparatus is similarly constructed but oriented more vertically, and the Mexican Golofa pizarro (233) has a horn which expands at its tip to form a three-pointed plate, concave downward, with the apexes of the plate pointing down and the inner surface covered with hairs. In Trypoxylus dichotomus (234), from Japan, the thoracic process is forked at the tip, and each of the two prongs is itself forked; in Chalcosoma atlas (235), from Sumatra, there are several processes which together form a sort of prehensile basket.

All these horn-like mechanisms of the rhinoceros beetles are excellently suited for grasping and clamping: the cephalic and thoracic processes move toward one another when the head is lifted. But it is not known whether this is their actual function. Only the males possess them; some have been demonstrated to serve as weapons, but others presumably have only a decorative function. In the dung beetles first mentioned, which shape dung into spheres, deposit eggs in them and roll them away to be buried, the horns may play a role in the transport of these materials.

Male stag beetles often have mandibles which have become enormously enlarged into "antlers" that can be closed together like forceps. Their inner margins bear teeth and at their tips are two- or three-pronged gripping structures. Examples include the Indian Cladognathus giraffa (236), the Siamese Hexarthrius deyrollei (237), and Neolamprima adolphinae from New Guinea (236a). Sometimes the mandibles are lengthened so that they look like scissors and are covered with spines, rather like the prehensile apparatus of the praying mantis (for example, in Chiasognathus granti (238) from New Guinea). When two male stag beetles engage in battle for a female, the rival can be siezed with these gigantic mandibles, lifted up and hurled away.

Plate 19 (Figs. 225–238) ▶

225. Double forceps of Damaster
226. Cephalic spur of Phanaeus imperator
227. Cephalic spur of Copris hispanicus
228. Thoracic spur of Heliocopris gigas
229. Thoracic spur of Enema pan
230. Thoracic process of Dynastes hercules
231. Pronotal horn of Dynastes neptunus
232. Pronotal horn of Golofa porteri
233. Pronotal horn of Golofa pizarro
234. Forked pronotal horn of Trypoxylus dichotomus
235. Prehensile basket of Chalcosoma atlas
236. Cladognathus giraffa
236a. Neolamprima adolphinae
237. Hexarthrius deyrollei
238. Chiasognathus granti

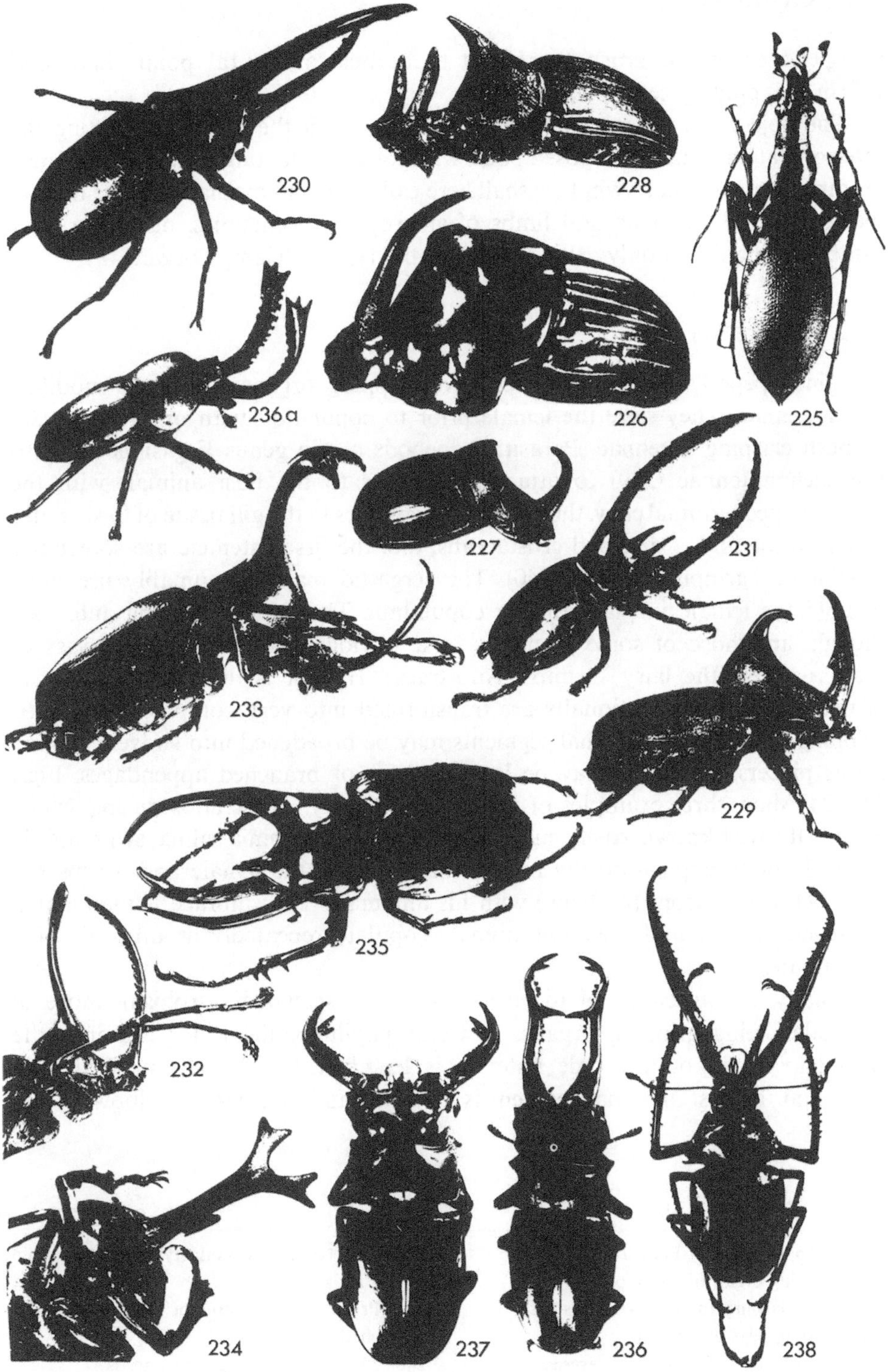
230
228
236 a
226
225
227
231
233
229
235
232
234
237
236
238

2.2.2 Grippers

Gripper-type organs grasp the object to be held at several points and hold it firm by applying centripetal pressure.

The object can be prevented from slipping out if the elements exerting the force operate in suitably opposed pairs, with rough contacting surfaces. Biological examples are abundant; what we shall here call "gripper" mechanisms are usually located on the antennae and limbs of arthropods. Sometimes, of course, they can be distinguished only arbitrarily from the typical "clamp" organs.

2.2.2.1 Nutcracker Type

Certain copepod (Crustacea) males have gripping organs which are modified first antennae. They sièze the female prior to copulation with one (Calanoida) or both clasping antennae. Parasitic copepods of the genus Ergasilus can also use such antennae (239) to attach themselves to the host animal; with the sickle-shaped terminal claw, they anchor themselves in the gill tissue of fresh-water fishes. In the higher decapod crustaceans, too, the first antennae are sometimes modified as gripping organs (240). The sergestid males presumably use them to hold the female in place during copulation. The oddest clasping antennae, though, are those of some members of a phylogenetically lower subclass of the Crustacea, the fairy shrimps (Anostraca). Here it is the second antennae of the male which occasionally are transformed into very complexly branched gripping organs. The terminal segments may be broadened into shovels or may act as pincers, and there may be long feathery or branched appendages. Figs. 241–243 show three examples of varying complexity. These anostracans, which include the well-known, easily cultured brine shrimp Artemia salina, swim upside down. Prior to copulation the male swims beneath the female and grasps her from below (i.e., from the back) with his antennae. The embrace is maintained for several hours, and then the animals copulate repeatedly at intervals of a few minutes.

Finally, the antennae of the fleas also show sexual dimorphism, those of the males being gripping organs. In Ceratophyllus gallinae, for example, the club-shaped end of the male antenna is lengthened and well supplied with hairs, and the first antennal segment is extended and bent into an elbow.

Plate 20 (Figs. 239–250) ▶

239. Clasping organ of Ergasilus
240. Clasping organ of a decapod
241–243. Clasping antennae of Anostraca
244. Clasping legs of Caprella
245. Clasping legs of Oncinopus aranea
246. Uroptychus glyphodactylus hanging from a stalk
247. Prehensile apparatus of Obidosus amplochelis
248–250. Prehensile legs in mantids

239

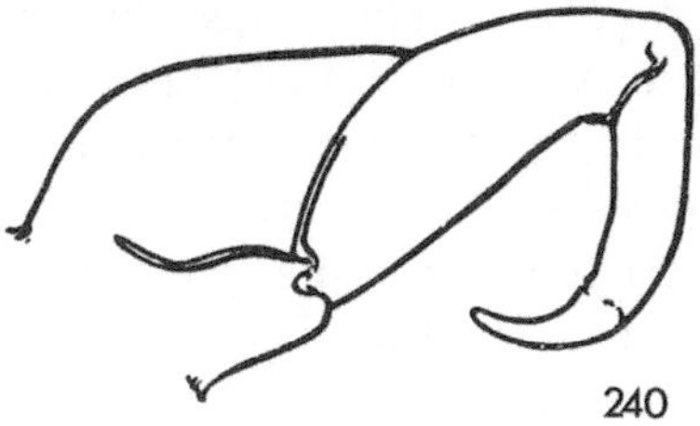
240

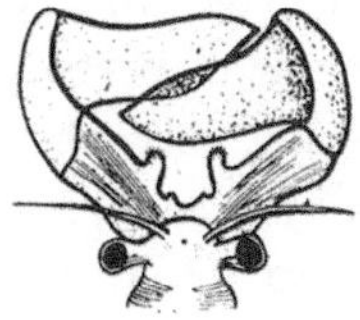
241

242

243

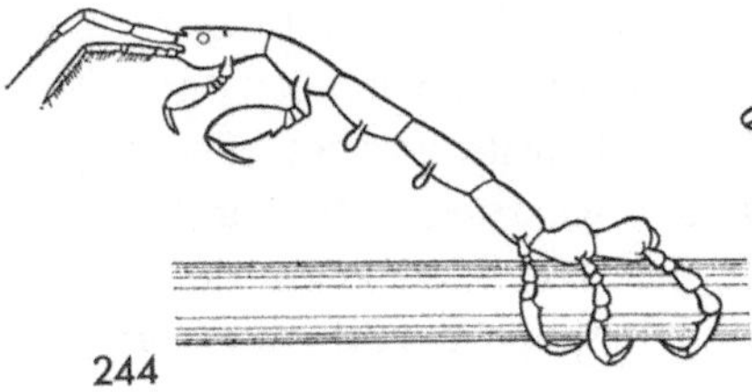
244

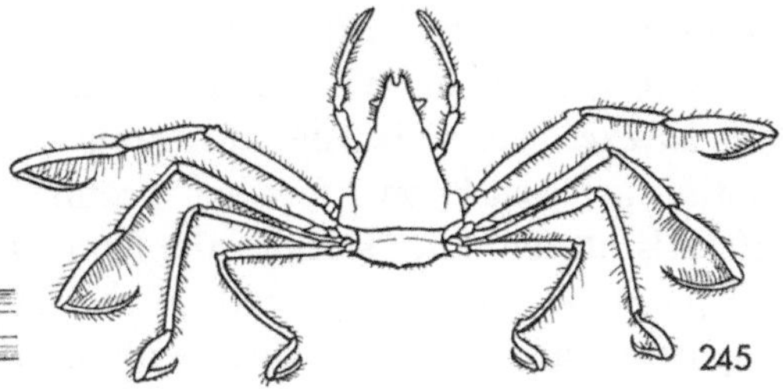
245

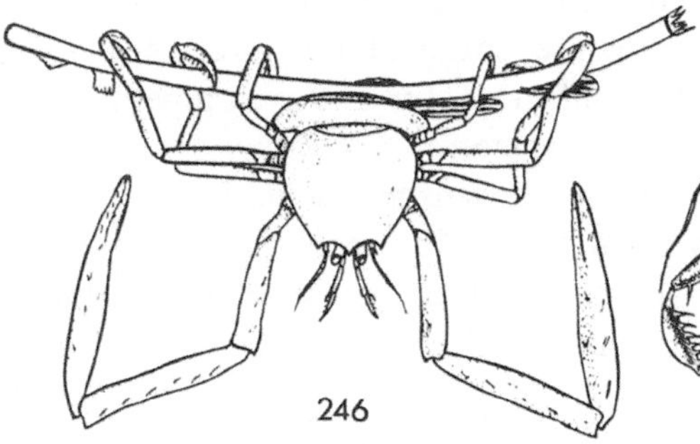
246

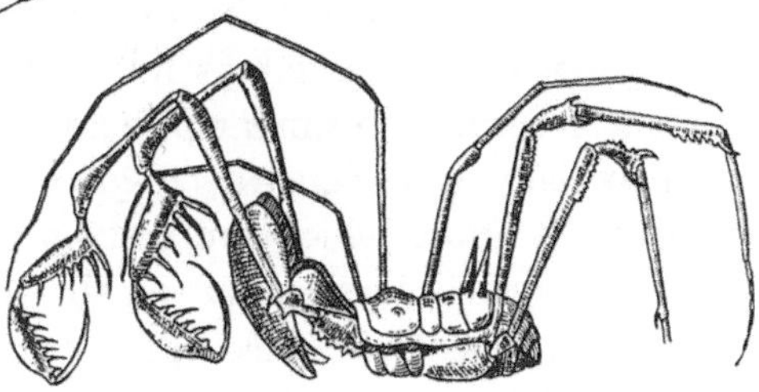
247

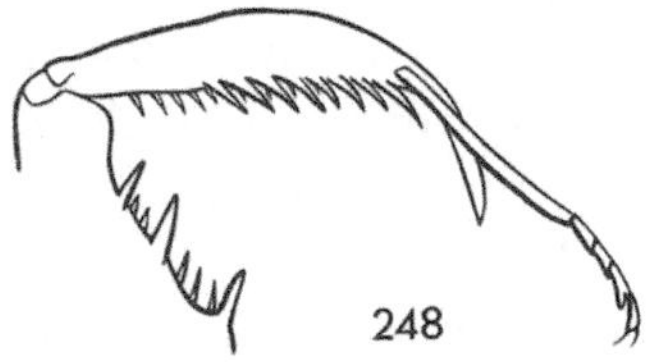
248

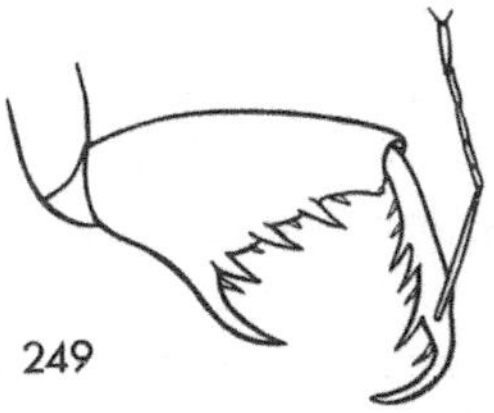
249

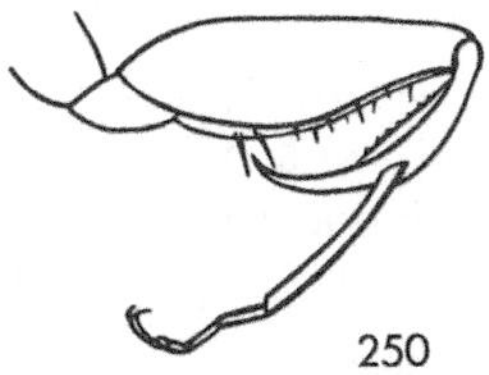
250

Prehensile legs are found in great variety among arthropods. The amphipod crustaceans called Caprellidae lurk in thickets of plants. With the two or three last pairs of pereiopods, they cling to the stems of plants or bryozoans, while the front part of the body is stretched out at an angle with the prehensile legs outspread (244). The legs operate in pairs, pressing toward one another so as to grasp the substrate in an extraordinarily reliable, well-fitting grip, similar to that of a split-sleeve clamp. Decapod Crustacea (Brachyura, Majidae) of the genus Oncinopus live in fields of seaweed off flat shores where the weeds are moved about by the waves. These animals have characteristic spiny gripping claws (245). But similar organs are found in deep-sea crabs, which live in absolutely still water. They use them to attach themselves to cylindrical octocoralline stalks and thus can hang head down, like laundry hung on a line with clothespins. Uroptychus glyphodactylus (246) and many other Galatheidae behave in this way.

The daddy longlegs Obidosus amplochelis, one of the many species belonging to the South American spider family Gonyleptidae, has very well developed terminal claws on its pedipalps. The two outermost segments bear numerous spines and have a hook at the end; they can be closed with great force and thus form an excellent prehensile apparatus (247). Among the insects, the praying mantises are well known for their prominent prehensile forelegs. The gripping surface is on the inner sides of the femur and tibia. Frequently the femur has a longitudinal groove into which the tibia fits like a pocket knife, and the gripping surface is always edged with strong hairs, spurs, and other processes. Figs. 248–250 show three types of the mantis gripping mechanism. It is used primarily to catch flies in a lightning-quick, preprogrammed grasping stroke. To this end, it is of course located far forward on the body, near the head, and it can be stretched far out because the coxae of the forelegs are very much elongated. Bone clamps, nutcrackers, alligator clips and even pipe wrenches operate according to similar principles of gripping.

Especially finely constructed gripping mechanisms are found in many mites, particularly those which are epizoic upon birds. The male Sulanyssus caput medusae lays his forelegs flat on the barbules of the flight feathers and pushes hooked appendages of his legs through them. These appendages now grasp the barbules from below and prevent slipping, in a way reminiscent of the locking mechanism of a trailer coupling. In Fig. 251 this connection is sketched. The wall-like structures to right and left are two adjacent barbs of the feather (cf. p. 76). The terminal segments of the legs of mites which live in fur are functionally similar to those of lice; these mites crawl along a single hair like a sloth on a branch, grasping the hair between the legs, which act as clamps. In Labidocarpus megalonyx, for example, the two most posterior pairs of legs are broadened distally like shovels to grip the hair, while on the most anterior legs there are long claws; these are bent into a semicircle at the tips, which, together with their opposed structures (252), operate exactly like crampons.

2.2.2.2 Four-jawed Grippers

All the above gripping mechanisms operate by the closure of two opposed parts. To make the attachment particularly firm, several pairs may be arranged in series, as in Caprella. But in nature there are also grippers which provide a firm attachment only if there is simultaneous closure of four contact elements. Male dragonflies use an apparatus consisting of four elements, the paired superior and inferior caudal appendages, to grasp the prothoracic region of the female (253); in the same way, special forceps with four claws can be used to pick up small objects. The "push-button-action" spreading clips of electronics function similarly.

2.2.3 Hooks

This category, too, is represented in nature by a variety of shapes and types. Hook mechanisms are often used to connect insect wings reversibly with one another. They can be used to attach the body of an animal to a substrate. In other cases projections of the body—which bear hooks or are themselves hook-shaped—are used to manipulate particles of food or other objects.

2.2.3.1 Hooks for Special Substrates

Coupling mechanisms which involve the principle of hooking one part into another are used to join the forewings and hindwings of bugs, lepidopterans, and hymenopterans. In these groups such mechanisms are quite varied in form; they are also found in other insect orders, though with a lesser degree of specialization. In particular, insects with two pairs of wings which beat at a high frequency, such as the hawk-moths, bees, and shield bugs, possess hook mechanisms in which the elements on the forewings and hindwings are finely matched. In this way the two wings on one side can be rigidly coupled, and thus beat in phase.

The butterflies and moths have two types of hook mechanisms. In the "jugate" type (254) a lobe on the posterior margin of the forewing *f* called the jugum *j* is pushed under the anterior margin of the hindwing *h*. Alternatively, in the "frenate" type (255) there may be a frenulum *fr* on the anterior margin of the hindwing; this consists of one or more bristles, and interlocks with a retinaculum *r* of the posterior margin of the forewing, which is usually in the form of a cushion of hairs but occasionally can be a rolled lobe of cuticle. The honeybee Apis mellifera has a single row of very fine hooks (hamuli) on the costal vein of the hindwing, all of which catch upon a slightly undercut ridge on the posterior margin of the forewing (256, 257). Pentatomid bugs of the genus Palomena couple the membranous hindwings to the hemielytra (forewings) by means of a snap mechanism which is secured against tensile stress. The hemielytra have a longitudinal groove on the posterior margin, along one edge of which there is a border of sturdy bristle-like hairs and on the opposite edge, a border composed of several rows of short stumpy bristles. These two borders suffice to hold a ridge (bent at right angles to the anterior margin of the hindwing and hammer-shaped in cross-section) securely in the hemielytral groove (258, 259).

A somewhat simpler, but in principle similar, hooking device is found in the bug Graphosoma italicum, which also belongs to the family Pentatomidae. Here the border of bristle stumps is replaced by a bevelled edge with overlapping plates (260–262). Not only the pentatomids, but also many winged heteropteran bugs possess such coupling devices; they have been found in such different families as the Cryptoceratidae, the Gerridae, and the Pyrrhocoridae. The Homoptera, too, have functionally similar devices. The hook mechanism of the European cicada Triecphora vulnerata (Cercopidae) is shown as a typical example in Fig. 263. A lobe of the hindwing margin (264) is drawn forward and bears several hook-shaped outgrowths which catch in the curled-up posterior margin of the forewing. Moreover, further out along the wings there is another projection pointing up and back (265), which fits into a correspondingly bent vein in the margin of the forewing.

In the cicadas, the posterior margin of the forewing is bent downward and catches in the upward-bent anterior margin of the hindwing. The Psyllidae (266), on the other hand, have a strong bristle, bent into a semicircle, on top of the anterior margin of the hindwing (268), which fits into one of the "embrasures" of the curled posterior margin of the forewing (267). The aphids (269), in the

Plate 21 (Figs. 251–273) ▶

251. Secured attachment of Sulanyssus caput medusae to a feather
252. Labidocarpus megalonyx
253. Fourfold gripping mechanism of a male dragonfly, attached to the female
254. Jugate type of wing coupling
255. Frenate type of wing coupling
256. Connection between the wings of Apis mellifera, seen from above
257. Detail of the joined edges (from 256)
258. Cross section of the connection between the wings of Palomena (Pentatomidae)
259. Surface view of the wing connection in Palomena
260. Groove fastener on the forewing of Graphosoma italicum, surface view
261. Anterior edge of the hindwing of Graphosoma italicum
262. Cross section through the joined wings (260 and 261)
263. Connection between the wings of Triecphora vulnerata, surface view. v, anterior fastener on hindwing; h, posterior fastener
264. Anterior fastening elements, enlarged from (263)
265. Posterior fastening elements, enlarged from (263)
266. Connection between the wings in the Psyllidae (Psylla buxi); cross section through the attached wing margins
267. Anterior margin of the hindwing, from (266)
268. Posterior margin of the forewing, from (266)
269. Connection between the wings in aphids (Drepanosiphum platanoides); cross section through the attached wing margins
270. Anterior margin of the hindwing (from 269)
271. Posterior margin of the forewing (from 269)

(272 and 273 omitted)

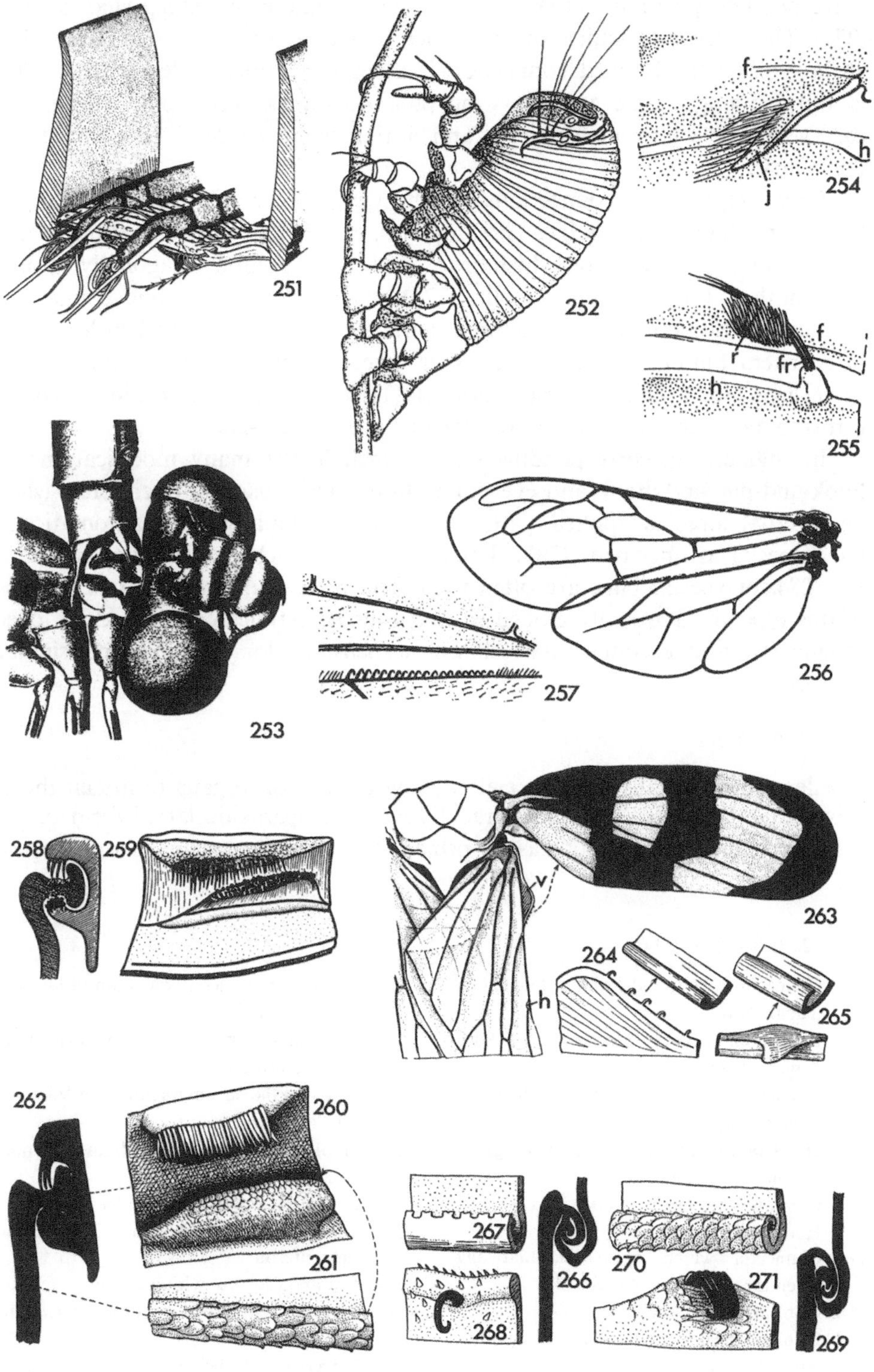
251
252
f
h
j
254
r
fr
f
h
255
253
257
256
258
259
v
263
264
h
265
262
260
261
267
266
270
271
268
269

corresponding position, actually have a whole bundle of semicircular bristles (271). The ridge into which these bristles fit has a rough scaly surface (270). In both cases the hooked connection is maintained only under tension; that is, only as long as the anterior wing pulls on the posterior one. This is the case when the wings are extended and during flight. At rest, the connections come apart.

Very simple hook connections for the wings are employed by the snake flies (Raphidiidae), caddis flies (Trichoptera), and Mecoptera. Sometimes these consist of only two lobes on which are a few bristles pointing diagonally toward one another. Raphidia maculicollis (274) has this sort of simple connecting mechanism. Sometimes a frenulum catches in a field of a few bristles, as in Drepanacra humilis (275), and in other cases the connection is limited to a single jugal lobe, as in Rhyacophila dorsalis (Trichoptera; 276). Some intermediate forms found in the Mecoptera are sketched in Figs. 277-280.

In engineering, corresponding devices include the many modifications of hook-and-pin and hook-and-eye connections. They usually open only when an external force is applied (281, 282), but occasionally open automatically by a special arrangement (283). Locking devices, such as the one shown in Fig. 284, or special pins, are often used. The biological coupling mechanisms of this type are frequently elongated and thus sometimes correspond to loose tongue-and-groove connections, sliding suspensions for curtains, and similar devices.

2.2.3.2 Anchors

Sessile animals usually have special clinging or fixation organs to attach themselves to a substrate. The oncomiracidium, an ectoparasitic larval form of the monogenean flukes, employs as its principal attachment organ an opisthaptor

Plate 22 (Figs. 274–299) ▶

274. Connection between the wings of Raphidia maculicollis
275. Connection between the wings of Drepanacra humilis
276. Connection between the wings of Rhyacophila dorsalis
277–280. Connections between the wings of various Mecoptera
281. Cam-operated latch clamp
282. Hook-and-pin closure
283. Connector which opens automatically when the stop (above) is struck
284. Hook with safety catch
285. Hook on the oncomyracidium of Acanthocotyle lobianclevi
286. Hook on the oncomyracidium of Diclidophora merlangi
287. Hook on the oncomyracidium of Acanthocotyle merlucci
288. Hook on the oncomyracidium of Microcotyle labracis
289. Hook on the oncomyracidium of Rajonchocotyle emarginata
290–294. Forms of medieval halberds
295 and 296. Hook mechanism of Tetraonchus monenteron
297. Crampon for use with a crane
298. Pincers in the proboscis of a turbellarian, opened
299. Pincers in a turbellarian proboscis, closed

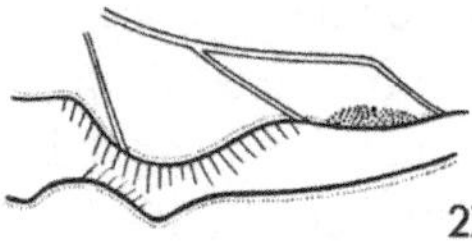
274

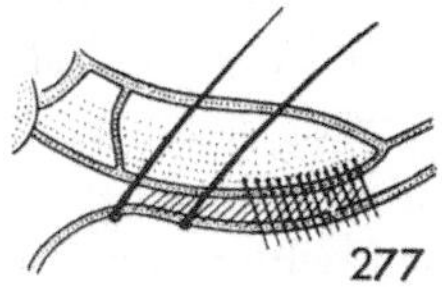
277

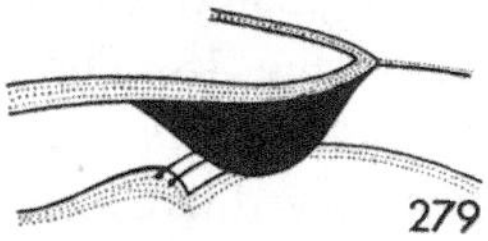
279

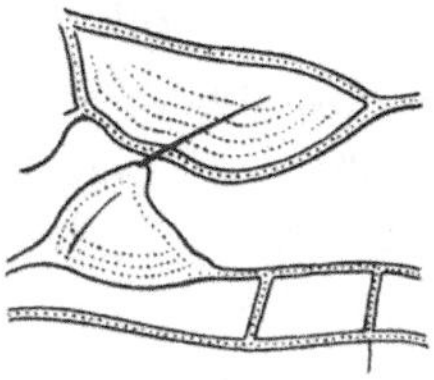
275

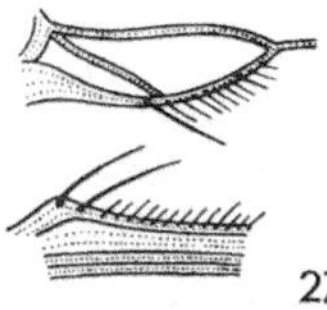
278

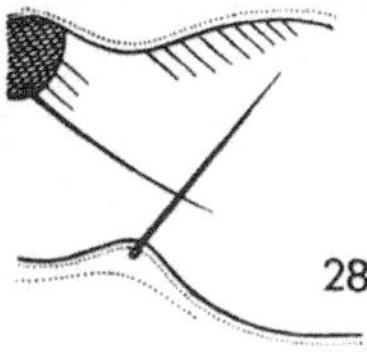
280

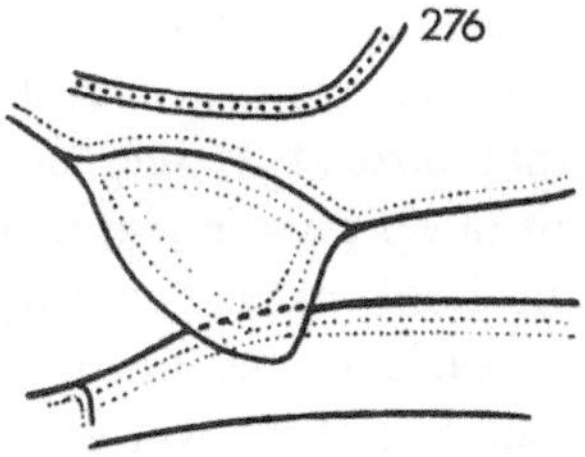
276

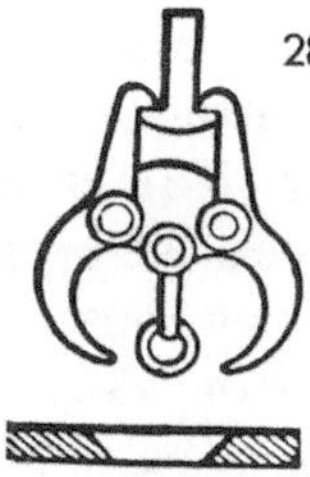
283

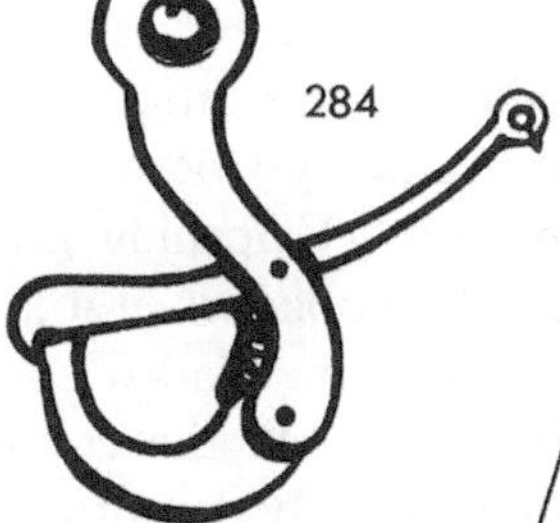
284

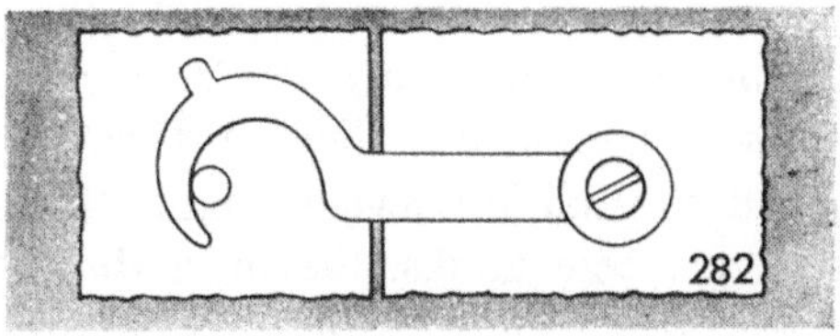
282

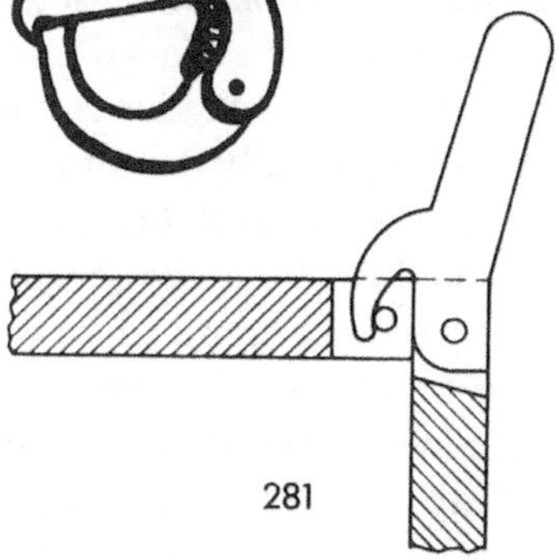
281

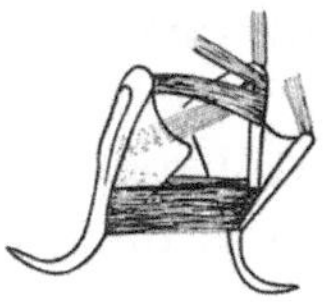
296

285

286

287

288

289

290

291

292

293

294

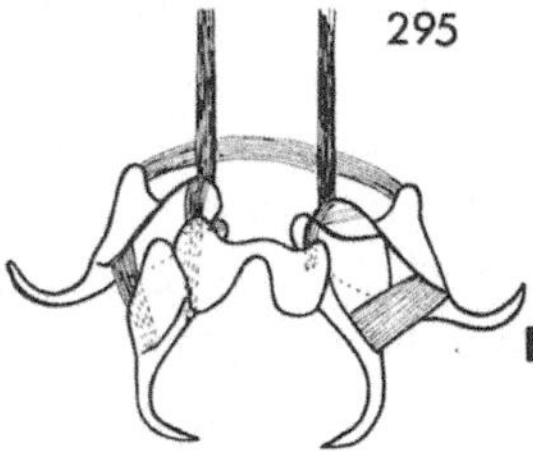
295

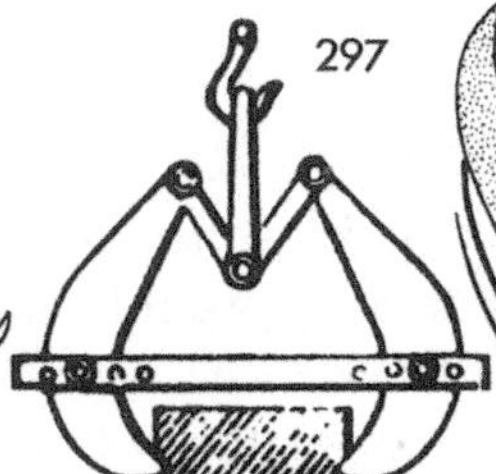
297

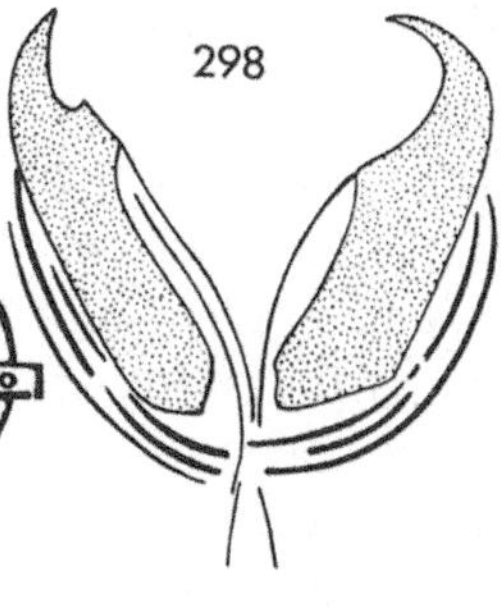
298

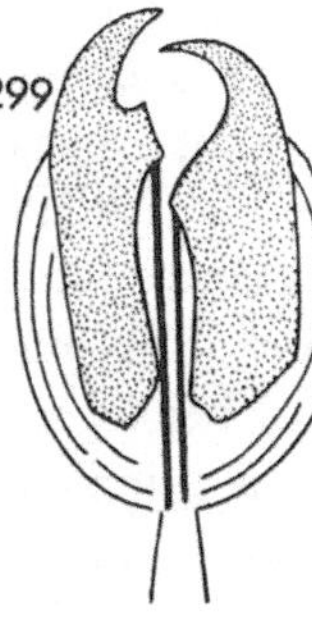
299

—a disc at the posterior end around which there is a double circle of hooklets, usually numbering 16. The individual hooklets differ in shape according to the species, but always consist of three parts: a shaft, a sharp cutting edge (more or less curved like a sickle), and a terminal process on one side (285–289). Like the halberds of the middle ages, which vary widely in form but always as modifications of a fundamental type (290–294), they are adapted both for penetration of tissue and subsequent anchoring there. These should also be compared with the different forms of the cyclophyllid hooks discussed below.

The adult form of the monogenean fluke Tetraonchus monenteron lives on the gills of the pike. It anchors itself there with four hooks, two dorsal and two ventral (295, 296), which can be moved by a special muscular apparatus. A transverse structure serves as an attachment point for some of the muscles and maintains the separation between the hooks. The actual operation of this mechanism has not yet been investigated. To judge from its form, it is related in function to a self-opening gripper (283), except that instead of the "stop" which opens the hooks when it is struck, the operation of opening is performed by contraction of the dorsal connecting muscles. The hook apparatus might also be compared with the crampons or "dogs" (297) on a crane, the points of which automatically dig into the piece being lifted because of its own weight. In this case, instead of the weight of the load, it is again muscular activity (the contraction of the two longitudinal muscles) which closes the hooks.

There are branchiuran Crustacea, commonly called "carp lice", which are ectoparasitic on fish. To attach themselves to the smooth slimy surface, these animals have powerful suckers (p. 95) and in addition several angled hooks at the base of the first pair of antennae; of these hooks, the most distal is especially long. The second antenna, which presses close to the first, is armed at the base with only one hook. The genus Dolops also wields a long clasping hook, on each member of the first pair of maxillae.

The larva of blood-sucking isopods of the genus Gnathia also lives on fishes. According to Kaestner, the larva "swims up to a fish, sinks the long, sickle-shaped terminal claw of each first pereiopod into a fish fin, like a boat-hook, and with these strong modified walking legs draws the mouth cone up to the skin of the fish. Then the mandibles, which can be moved only slightly, are pushed into the skin until their barbs hook in place like harpoons. During the entire act of sucking, the isopod remains attached to the host with only these two pairs of appendages."

The turbellarian worms belonging to the Kalyptorhynchia have a conical proboscis which in some cases has a pair of hooks at the tip; these can be stuck like pincers into a prey animal by powerful muscles (298, 299). Many bird beaks, as well as the mouth parts of some cephalopods, operate on the same principle. There are also instances of two rows of hooks which work against one another: the pectinate pedicellariae of some sea stars consist of 5–6 bent hooks in parallel in each row, and the two rows can be pulled together.

As a rule, the tongues of birds are horny; the lateral edges at the tip are sometimes equipped with keratinized teeth which point back at an angle. These teeth act like the barbs on an arrowhead; prey stabbed with the tongue cannot work loose. In Figs. 300 to 303, the tongues of the peacock, heron, dipper, and woodpecker are sketched, along with some arrowheads (304–306) used by Brazilian Indians.

In the tropical rain forest there are many climbing plants. The most successful of these are the climbing woody plants which hook themselves into their host plants with thorns or spines. The fern Gleichenia linearis climbs many meters in this way, and the rattan palms of the genus Plectocomia grow as long as 200 to 300 meters. Their pinnate leaves end in terminal flagella several meters long. Like the other parts of the leaf and the leaf sheath which extends down the stem, they are covered with barbs as hard as metal (307–309). These anchor the plant in the leafy roof of the jungle. Typical representatives of the twining plants may carry additional strong climbing hooks on their stems, with which they anchor themselves loosely in rough or powdery moldering substrates; the genus Uncaria (from *uncus*, "hook") (310) belongs to this group. There are also examples among the European flora. The stalk of the hop plant is equipped with longitudinal rows of anchor-like climbing hooks, the runner bean has climbing hairs in the shape of crampons, and many fruits and seeds carry barbs which aid their distribution (320). In technology the best-known examples of such devices are the various forms of anchors (311–314) and fish-hooks, which serve so well to catch and hold.

2.2.3.3 Multiple-hook Devices

In the examples cited thus far, each mechanism has been composed of only a few gripping hooks, which often are arranged in groups or in opposed pairs. Some grasping systems, though, operate with a large number of hooks. These are found in arrays upon special anchor organs. Tapeworms of the cestode order Trypanorhyncha (= Tetrarhynchoidea) have been given the latter name because of the four proboscises on the scolex. These proboscises can be completely withdrawn into four strong proboscis sheaths. When everted they are hollow tubes armed with hooks (315) which can effectively anchor themselves in the gut wall: with one exception all representatives of this order of tapeworms live in the gut of selachians (sharks and rays). The hooks on the trypanorhynch proboscis have been described as of three sorts: primary, accessory, and localized. Primary hooks, also called macrohooks, surround the proboscis in typical half-rings, so that there is room between them for additional rows of accessory hooks, or microhooks. Localized hooks are those which are found only at certain places on the proboscis. The hooks may also be distinguished according to size and shape—giant hooks, hooklets, microhooklets, and hook chains (316, 317).

Technological devices in the form of rods or tubes, which are designed to be anchored in various substrates so that they cannot be pulled out, are also provided with functional analogs of barbs: the ridges on nails for softwood (318), circumferrential bulges in plastic wall plugs (319), and the fine barbs on the broaches used by dentists to ream out the nerve from a root canal. Among plants, too, there are closely related anchoring mechanisms. The fruits of Bidens connatus, for example, have two to four pointed apical processes equipped with many tiny barbs. Other forms, shown in Figs. 320–324, are found on the fruits of aquatic plants.

In members of another tapeworm order, the Cyclophyllidea, the apex of the scolex forms a muscular rostellum on which strong hooks are arranged, either as a central, precisely circular wreath or in two circlets which overlap, with the hooks staggered. The species Hymenolepis macracanthus from the gut of the merganser and Cysticercus pisiformis from the gut of the hare are typical examples of these two groups; they are illustrated in Figs. 325 and 326 (cf. also Fig. 327). The arrangement and manner of movement of the hooks in Hymenolepis correspond to those of the claws in certain hay forks (328), which—as in the tapeworm—can be brought together when a pull is applied and thus grasp a certain amount of the substrate. Some earth-lifting machinery, like the commercial multiple-claw crane attachment shown in Fig. 329, operates on the same principle. So do the concentrically arranged and centripetally sprung stilettos of the fish-spears used by the Amazon Indians (330) and the circularly arranged barbs of the fruits of the agrimony, Agrimonia eupatoria (331). Other mechanisms belonging to this group are the tridentate pedicellariae (cleaning and prehensile organs) (332) and the six-element "Aristotle's lantern" (feeding organ) of the sea urchins, the four- to fourteen-element grasping-hook arrangement on the head of the arrow worms (333), and, in a way, the many-armed prehensile apparatus of the hydroid polyps and sea anemones. All these multiple-hook

Plate 23 (Figs. 300–326) ▶

300. Tongue of a peacock
301. Tongue of a heron
302. Tongue of a dipper
303. Tongue of a woodpecker
304–306. Arrowheads used by Brazilian Indians
307–309. Thorns in three different Plectocomia species
310. Climbing hooks in Uncaria
311–314. Types of anchors
315. Structure of the proboscises inside a trypanorhynch cestode
316. Types of hooks on the proboscis of a Grillotia erinaceus larva
317. Hooks on the proboscis of a trypanorhynch cestode
318. Nail for softwood
319. Plastic wall plug
320. Barbs on the fruit of Ceratophyllum submersum
321 and 322. Barbs on the fruit of Trapa natans
323. Barbs on the fruit of Trapella sinensis
324. Barbs on the fruit of Zanichellia palustris
325. Circlet of hooks in Hymenolepis macracanthus
326. Circlet of hooks in Cysticercus pisiformis

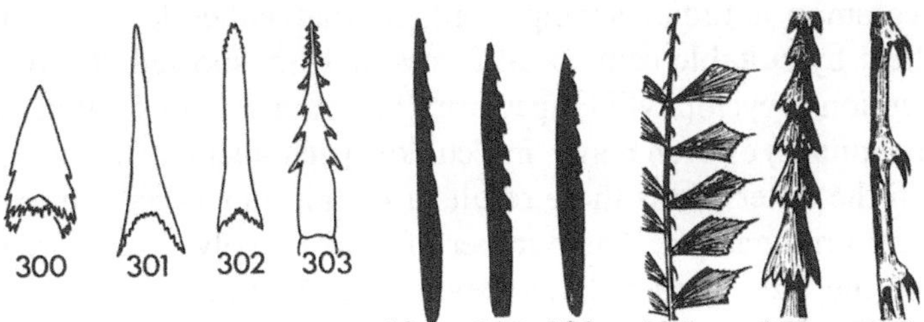

300 301 302 303 304 305 306 307 308 309

315

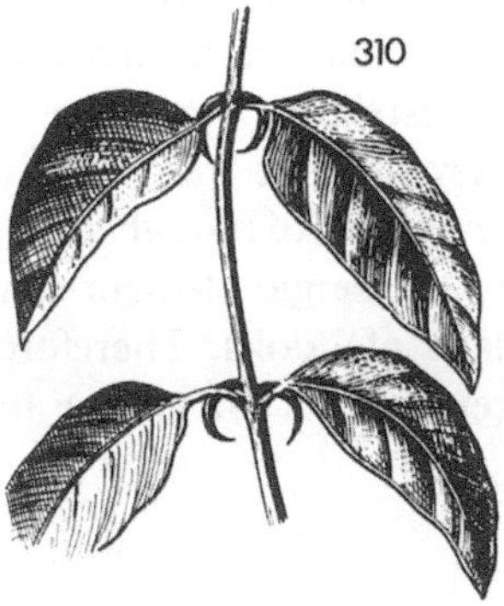

310

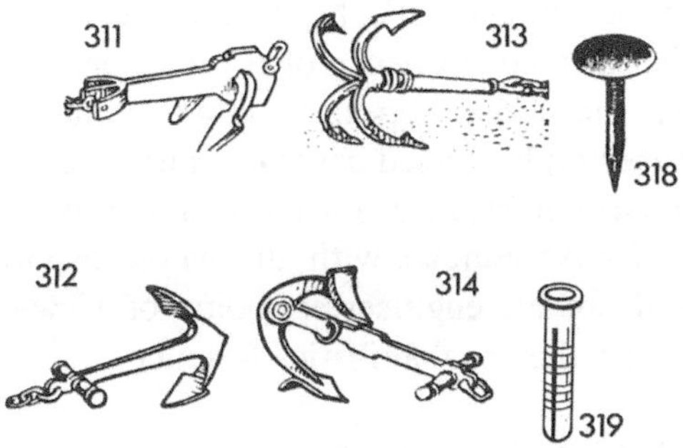

311 312 313 314 318 319

316

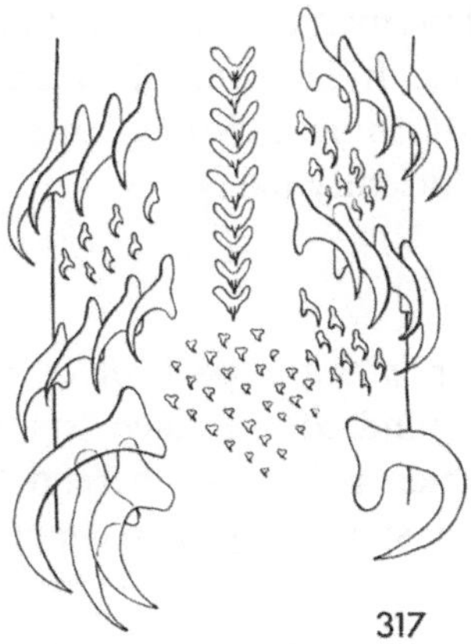

317

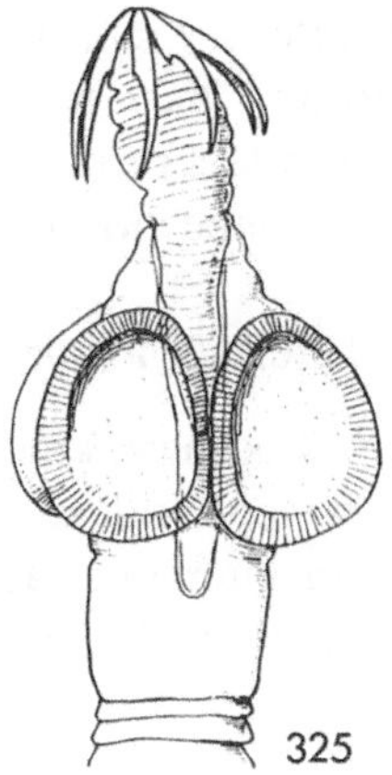

325

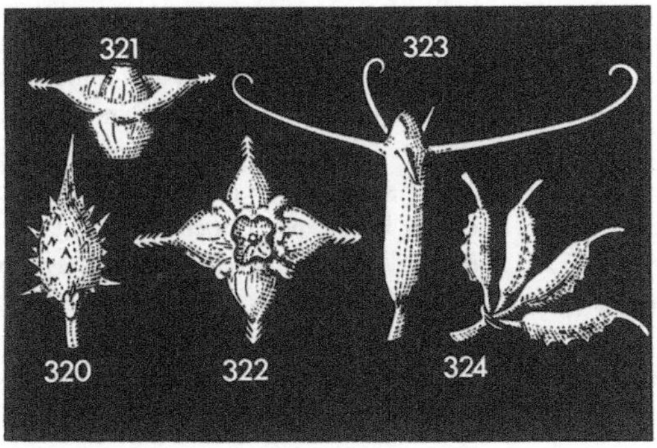

326

graspers have in common a radial arrangement, so that either loose or firm materials can be held by suitable centripetal forces and transported effectively.

In the above-mentioned cyclophyllid tapeworms one occasionally finds several (up to 16 in the Dipylidiinae) or even many (in Echinorhynchotaenia tritesticulata) circlets of hooks; in the latter forms there results a certain functional similarity to the trypanorhynch arrangement. The number of hooks involved can amount to from 8 to 1000, the size of each hook being 5 to 450 μm, and the form, again, is species-specific (334–349). All these cuticular structures serve to anchor the worm in the epithelium of the gut. For this purpose they are sometimes moved by special antagonistic hook muscles but usually the whole array is moved together as a consequence of contractions of the rostellum.

Thus when the hooks appear in large numbers, they are not individually movable, but are simply pressed passively into chance irrgularities of the substrate. Here, attachment depends not on the characteristics of the single element but rather upon effective contact with the whole assemblage of hooks. Therefore these forms—from an engineering point of view—represent a transition to the next category, the "probabilistic" fasteners.

2.2.4 Probabilistic Fasteners

This category can also be described as "random hooking" or, after the chief subgroup, "burr type" devices. It is required of any one hook only that it have a certain probability of attachment. With a large ensemble of such elements, a good connection is then sufficiently likely.

2.2.4.1 Burr Principle

The best-known kind of burr is the large fruit of Arctium lappa (350). The fruit is spherical and bears hundreds of long "spines" set at approximately uniform intervals, each of which is bent at the end to form a strong pointed barb. The shorter barbs of the spine ring of Agrimonia eupatoria (351) appear

Plate 24 (Figs. 327–360) ▶

327. Circlet of hooks in Taenia pisiformis
328. Hay fork
329. "Demag" multiple-claw grab
330. Fishing spear of Amazon Indians
331. Fruit of Agrimonia eupatoria
332. Tridentate pedicellaria of a sea urchin
333. Grasping hooks on the head of a chaetognath
334–349. Various types of hooks in the Cyclophyllidea
350. Barbed spine of the fruit of Arctium lappa
351. Barbed spine of the fruit of Agrimonia spec.
352. Barbed spine of the fruit of Cynoglossum spec.
353. Barbed spine of the fruit of Bidens spec.
354 and 355. Burr-type "zipper" fastening
356 and 357. Anchor filaments on the egg of Bdellostoma stouti
358. Single filament from (356–7)
359. Single filament from (356–7)
360. Hooks on the statoblast in the tissue of an ectoproct

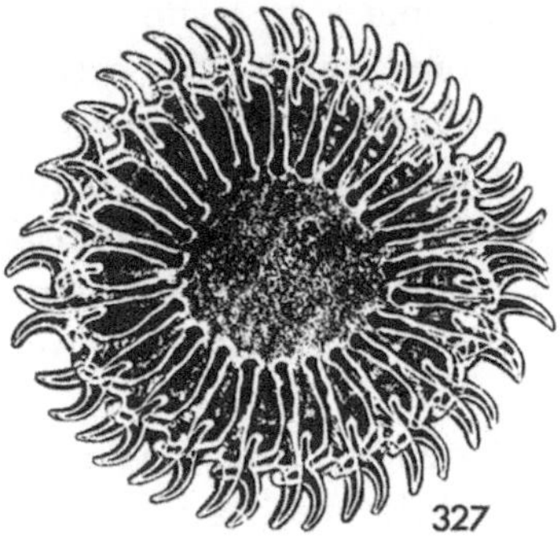
327

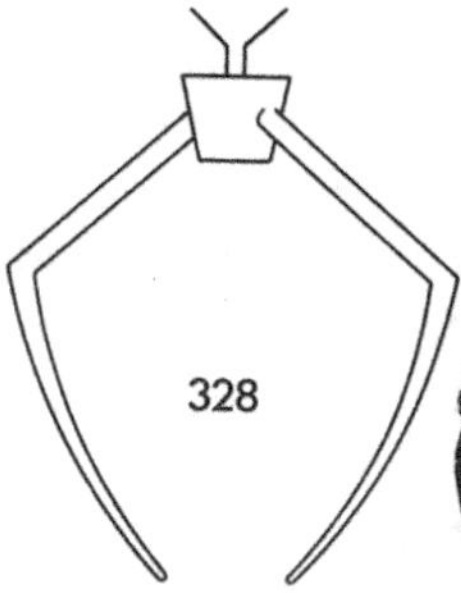
328

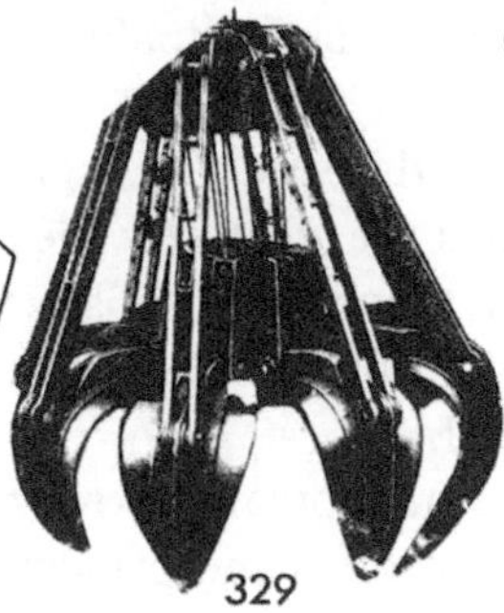
329

330

331

332

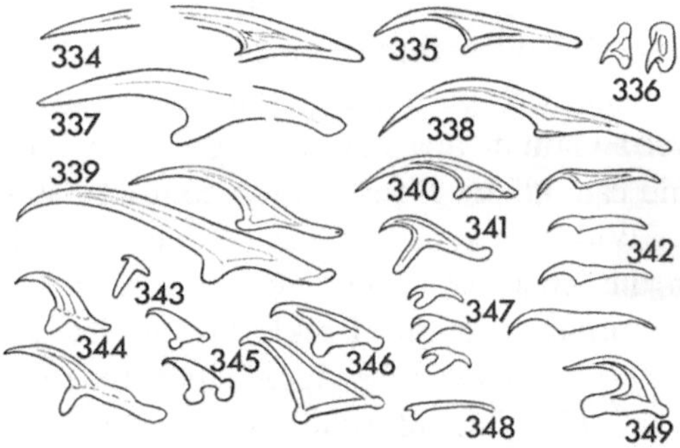
334 335 336 337 338 339 340 341 342 343 344 345 346 347 348 349

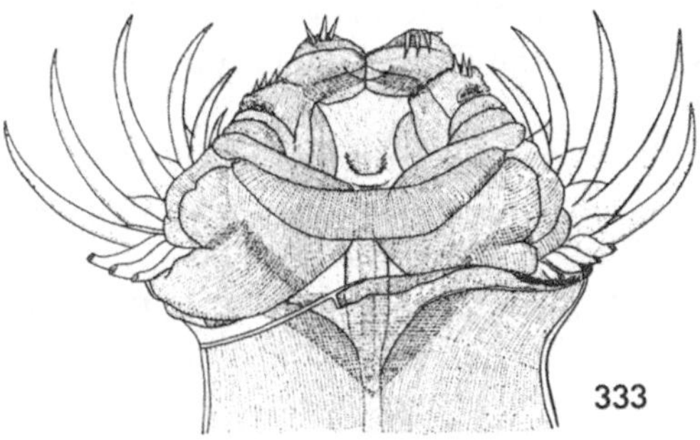
333

350 351 352 353

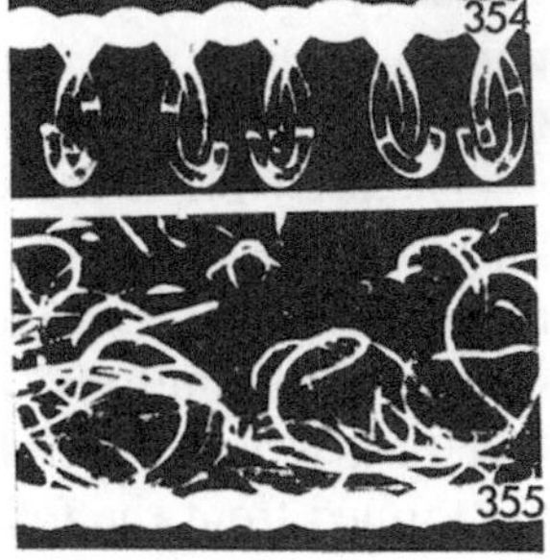
354 355

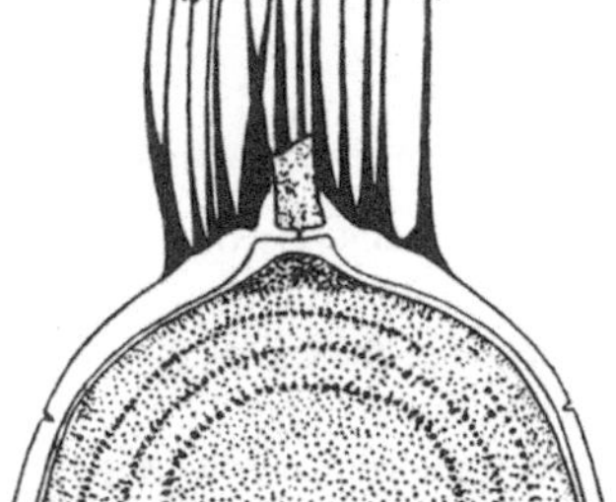
356

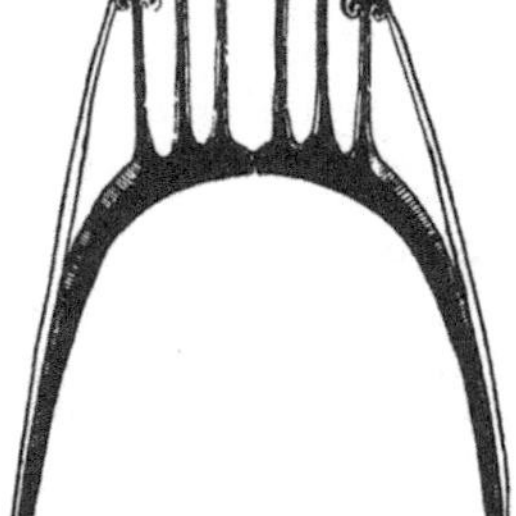
357

358

359

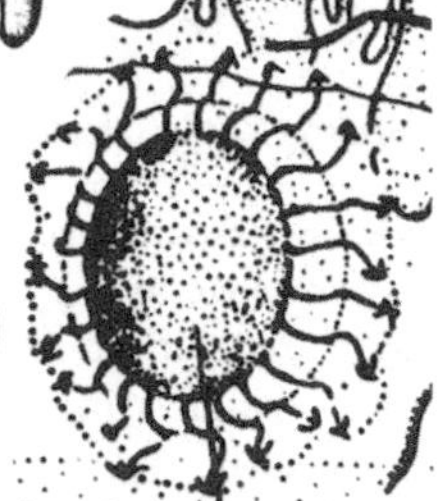
360

similar under magnification, whereas the tips of the spines of Cynoglossum have the form of a double anchor (352). All of these "burr fruits" (353) become caught quite readily in the pelts of animals and the clothing of humans. In this way the seeds are spread over a wide area. The attachement is better, the fluffier the substrate. The barbs of the fruit anchor themselves in a probabilistic manner among the substrate fibers; some fail to attach if they happen not to touch a fiber, but it suffices if a certain fraction of the total becomes firmly anchored.

Fasteners for fabric, known commonly as "Velcro" fasteners, have been devised which use the same principle. Of the two pieces of fabric one bears a "hook tape" with strong broken loops of perlon and the other, a "nap tape" of perlon threads which are only half as thick and are matted together like cotton wadding (354, 355). When the two tapes are pressed together they become securely attached; a strong jerk pulls them apart again. In a similar way, the leaves of the sticky weed Galium aparine are provided with thousands of extremely fine barbs, and can adhere closely to the contours of felt-like or porous surfaces.

With this type of fastener one can, for example, achieve small alterations in the length of a belt. This is a considerable advantage in comparison with the principle of the buckle and eyelet, which allows the belt to be adjusted only in discrete, rather large steps. One can now buy a type of photographer's case which has adjustable compartments—the dividers are held in place by the principle of probabilistic fastening so that they can be fixed in any position. The compartments can easily be altered to accommodate a new item of photographic equipment. This approach may yet play a significant role in the technology of packaging.

The principle of probabilistic fastening is not at all uncommon in the animal kingdom. The eggs of the European fish Bdellostoma stouti (Myxinidae, Cyclostomata), which can be up to 20 mm long, exhibit at both ends a bundle of hornlike threads with "buttons" at the tip (356, 357). Each of these little buttons has four fine, well-developed barbs (358, 359). These so-called "anchor filaments" are formed from the follicular epithelium. In the water they extend and catch in the hooks of nearby eggs so that the whole egg-mass tends to stay together in bunches or clumps.

Plate 25 (Figs. 361–371) ►

361–364. Examples of eggs in different species of Macrobiotus
365. Barbs on the opisthaptor of Acanthocotyle
366. Armature of the proboscis of an acanthocephalan
367. Armature of the proboscis of Centrorhynus aluconis
368. Armature of the proboscis of Acanthocephalus dirus
369. Armature of the proboscis of Neoechinorhynchus crassus
370. Part of a nematomorph larva with inverted proboscis
371. Crest of hooks on a kinorhynch

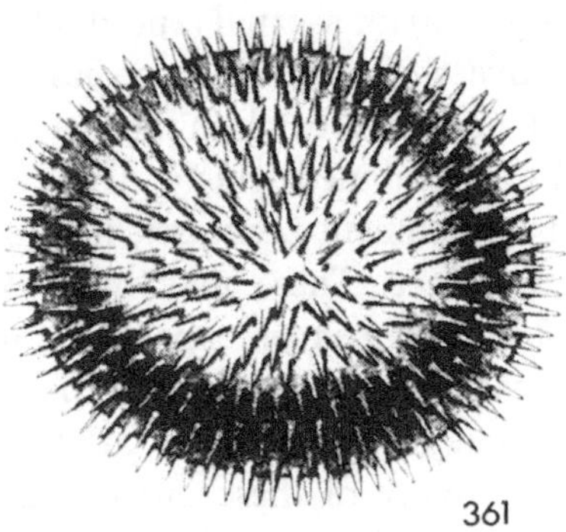

361

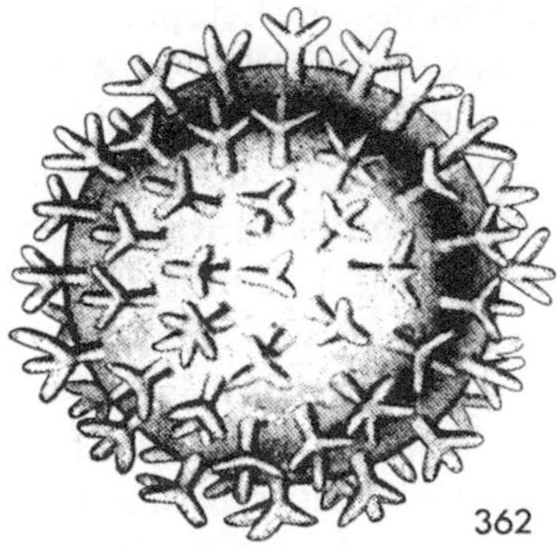

362

366

363

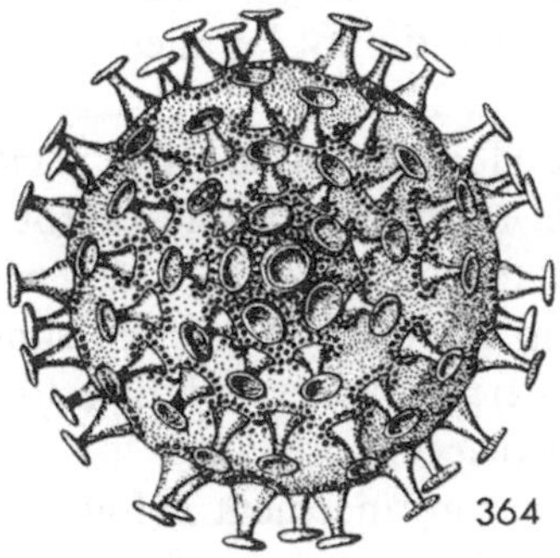

364

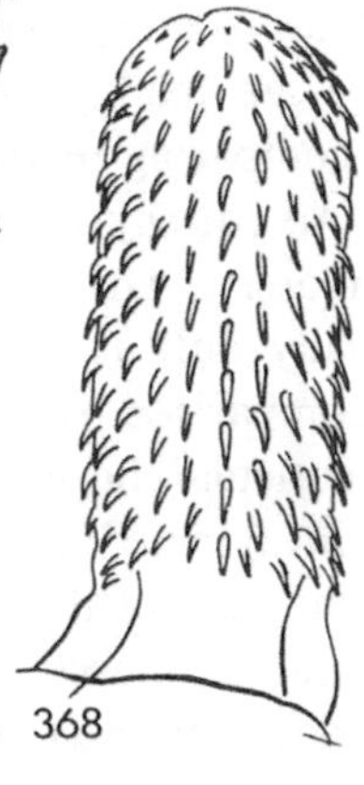

368

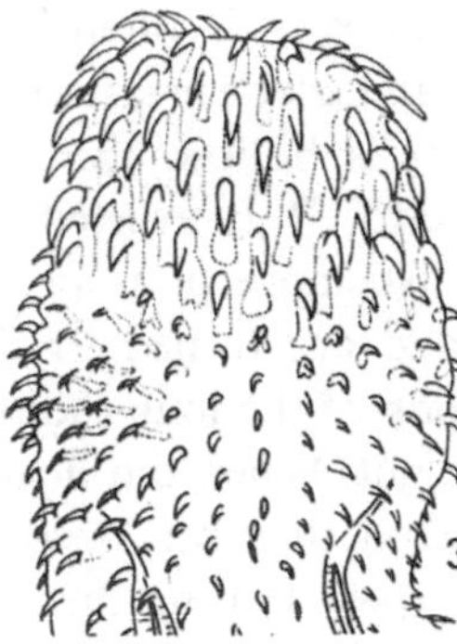

367

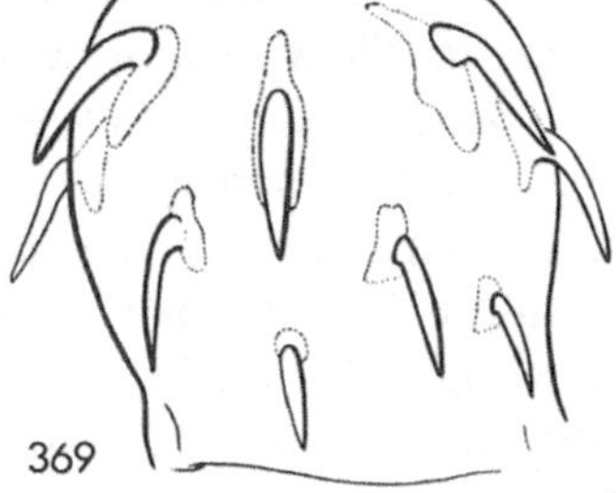

369

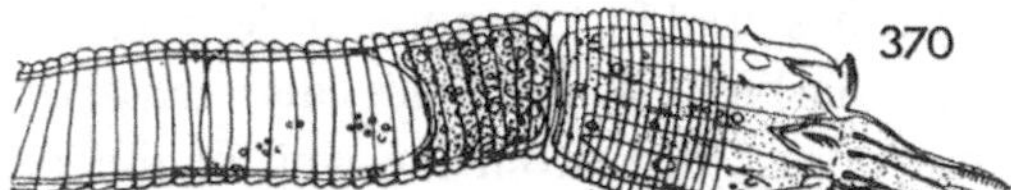

370

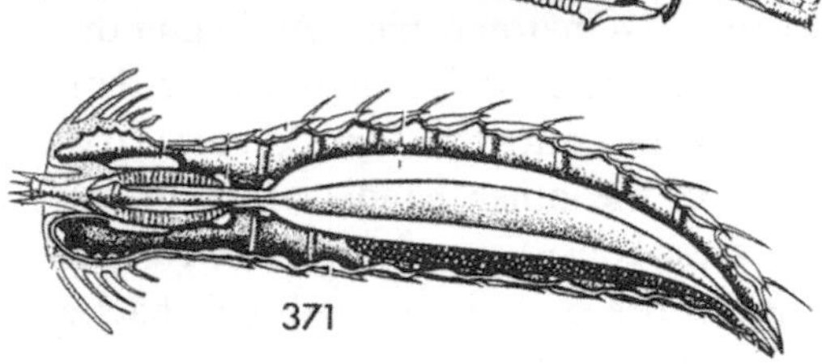

371

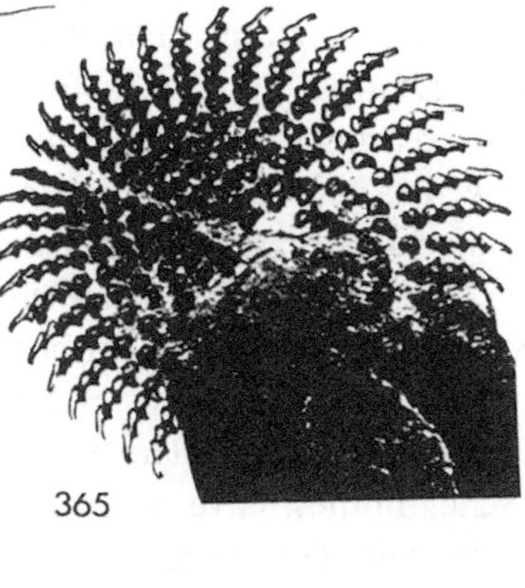

365

The statoblasts produced by Ectoprocta (= Bryozoa) for asexual reproduction consist of a cell mass surrounded by a protective shell. In some species the shell bears a circle of tiny, anchor-shaped hooks (360) which catch in the feathers of water birds. In this manner the ectoprocts are distributed widely. The eggs of some tardigrades are provided with chitinous attachment processes, not just at one end or in a circle but over the entire surface. These may have the form of spines or clubs, or the ends may be three-pronged or expanded into flat dish-like shapes. Figs. 361–364 show some eggs from species of Macrobiotus. Each of these eggs is laid individually. They become anchored in an arbitrary orientation in tiny irregularities of the substrate; many of these animals live in cushions of moss on rocks and roofs, through which the rain often produces a strong current. The anchoring mechanism keeps the eggs from being washed away. Other tardigrades do not lay their eggs individually but rather in masses, which remain together. The eggs of these species are always smooth!

The monogenean fluke Acanthocotyle pacifica lives on rays. To anchor itself to the smooth skin of the fish it uses a round opisthaptor which is covered with a remarkably regular array of radial rows of strong barbs (365). Other species of this group of worms have more simply constructed attachment discs. One can readily interpret their evolution toward this most complicated type by comparing related species.

An equally regular arrangement is found in the longitudinal rows or rings of massive barbs on the proboscis of acanthocephalan worms. These animals, usually only millimeters long, are gut parasites. The proboscis is often tiny in comparison to the whole body, but because of its strong and effective armature of cuticular hooks it is an excellent anchoring organ. It is hollow and can be invaginated at the tip so that the hooks point inward. It is everted by an increase of the internal pressure, sometimes with the assistance of a special protractor muscle. As the teeth appear at the margin and are turned to the outside they catch in the villi of the gut. Then, by contraction of the longitudinal muscles of the body, the worm pulls itself forward in the gut. When the hooks have been driven into the gut wall as far as the submucosa they can become so firmly anchored that the worm can no longer be expelled from the gut by peristalsis. Figs. 366–369 show some types of proboscis armature in order of decreasing complexity. The round brushes used for cleaning bottles and pipes, as well as wire brushes, are similar after a fashion; their bristles catch in particles of dirt in a probabilistic manner and loosen them from the wall of the pipe.

The larvae of the horsehair worms (Nematomorpha, Aschelminthes = Nemathelminthes) have proboscises with stylets and with rings of recurved bristles at the base (370). They use these to burrow into the thin-skinned larvae of fresh-water insects. The Kinorhyncha, on the other hand, a marine class of Aschelminthes, have on their heads a crest consisting of circles of rigid hooks or "scalids" (371); occasionally even the entire body is covered with backward-pointing bristles. Taken together, these act as a mushroom-shaped mud anchor

(372). The animals anchor themselves while crawling in the sandy floor of the ocean, as Kaestner describes it, "by everting the hook-bearing 'head' and pushing out the hooks, attaching themselves in this way to the ground and then shortening the body so as to pull it along behind. Then the body is extended again by means of the dorsoventral musculature—the anterior end is pushed forward in the direction the animal is crawling, by about 1/3 of the body length." Movement is thus achieved by the "inchworm" principle.

Many of the echiuroid worms dig passages in the sandy ocean floor. They crawl in these by means of peristaltic longitudinal waves of the body, as do the kinorhynchs. A circle of bristles on the posterior end, which catches in the wall of the tube, acts as a mud anchor (372). Phoronis anchors itself in the sand by the swollen end of the body (373). The mite Hyalomma cypriacum drills slit-like holes with its gnathosoma and then pushes its hypostome into the wound, anchoring itself there with the well-developed barbed plate of the hypostome (374). Finally, it is common knowledge that the tongues of some vertebrates, for example those of the cat (375) and cow, are covered with fields of recurved horny teeth. The tongue thus serves very well as a brush or curry-comb when the pelt is cleaned.

2.2.4.2 Comb Principle

When one combs rough hair with a very fine comb it catches and soon cannot be moved, because individual hairs running at angles to one another entwine tightly at the base of the comb. This principle is used with surprisingly little modification by various groups of ectoparasitic insects living in the fur of vertebrates. Their combs or "ctenidia" consist of one or more rows of closely and uniformly spaced bristles. Such combs have evolved independently in several ectoparasites, including the fleas (Siphonaptera), the bat flies (Nycteribiidae), the bat bugs (Polyctenidae), and the "beaver beetle" Platypsyllus castoris. The distance l between two "teeth" of a given comb is related with astonishing precision to the hair diameter d of the corresponding host animal; the graph of the function $l(d)$ shows that the relationship is a linear proportionality (376). This distance l, at the tips of the bristles, is always about 1.75 times the diameter d of the host hair, so that the hairs readily enter the comb. Toward the base of the comb the gap between teeth narrows nearly to zero. Somewhere in the middle, therefore, the hair will be caught by slight irregularities which are always present. Furthermore, the fine matted hairs lying below the actual fur can also become entangled in the teeth of the comb. The ctenidia are always slanted backward so as not to hamper the locomotion of the ectoparasite, which always runs forward. But if the host animal combs and grooms its pelt with tongue and snout, the parasites take up a characteristic diagonal position with the head down. Now they can only be pulled out backward, which the ctenidium very effectively opposes. This resistance can be felt directly if one does the experiment of gluing a dog flea onto a little stick and pulling it backward through the fur.

2.2.4.3 Feather Principle

A particularly finely developed kind of probabilistic fastening is found in the feathers of birds. It is well known that one can break the vane up into segments by pulling on it tangentially; stroking it lightly with the finger is enough to close the vane again and reestablish the uninterrupted surface. Birds regularly eliminate such slits after a flight by stroking through the feathers several times with the closed bill or even (as with hummingbirds, Fig. 377) pulling single feathers rapidly through the opened bill.

The vane of a typical contour feather consists of a number of side branches called barbs which are aligned in parallel and at an angle to the shaft, or rachis. The barbs carry on either side a similar array of second-order side branches called barbules (*b* and *h* in 378 and 381). In most birds the row of barbules on the side of the barb toward the tip of the feather bears barbicels, very fine backward-pointing hooks. These barbules may therefore be referred to as "hook barbules" (379; *h* in 378, 381). The barbules of the other side are without teeth and are usually curved sharply, so that an apt name for them is "bow barbules" (380; *b* in 378, 381). The row of hook barbules on a barb becomes anchored in the row of bow barbules of the adjacent barb, with no particular matching of elements required. With all of the barbs hooked together, the vane presents a uniform closed surface.

As a result of the favorable orientation of barbules the barbs attach almost automatically; a light stroke of the bill is enough to accomplish this apparently complex task (377). Strong local pressures, which would destroy a solid vane, here simply disconnect the probabilistic attachments, allowing the vane to divide

Plate 26 (Figs. 372–385) ▶

372. Mushroom anchor, for use in muddy bottoms
373. "Mushroom anchor" of Phoronis spec.
374. Barbed plate on the gnathosoma of Hyalomma cypriacum
375. Horny teeth on the cat tongue
376. Linear relationship between interbristle distance *l* of the ctenidia and diameter *d* of the hair of the host. Open circle, Nycteribia biarticulata; filled circles, 15 species of flea; square, Platypsyllus castoris
377. Smoothing of the feathers by the hummingbird Aglaiocercus kingi, ♂; photographed with an electronic flash device
378. Diagram of the hook (h) and bow (b) barbules of a feather
379. Single hook barbule
380. Single bow barbule
381. Schematic drawing of the feather structure: r, barb (or ramus); h, hook barbule; b, bow barbule
382. Photograph of a pheasant feather, showing hook barbules (h) and bow barbules (b)
383. Foot of the gecko Phelsuma
384. Foot of the gecko Hemidactylus
385. Foot of the gecko Ptyodactylus

373

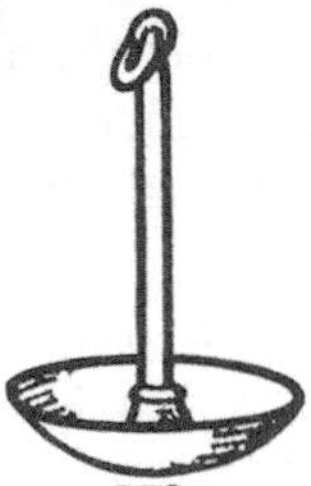
372

374

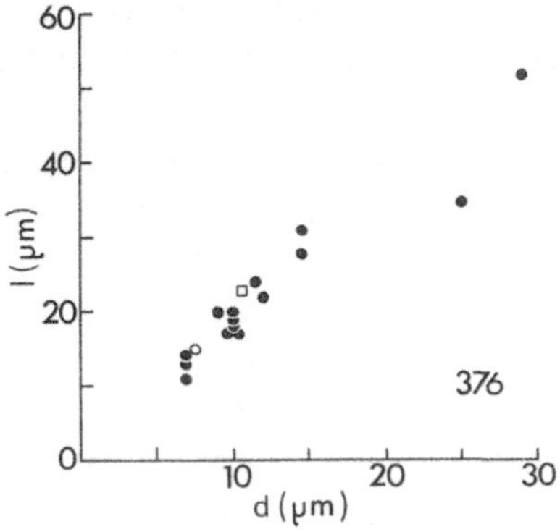

376

375

383

384

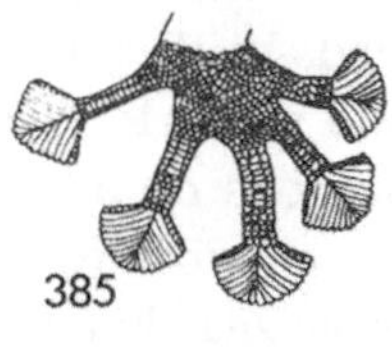
385

377

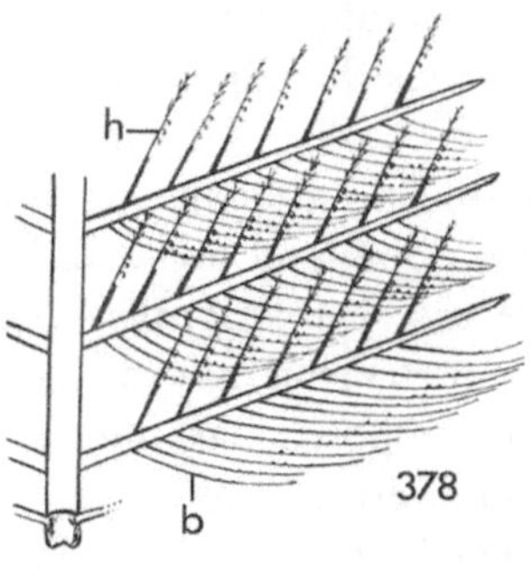

378

379

380

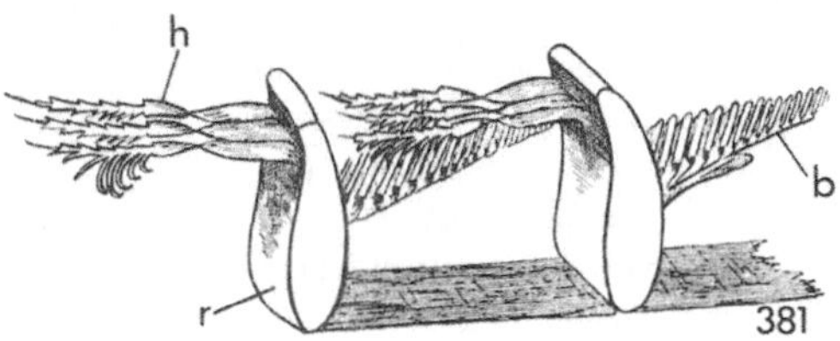

381

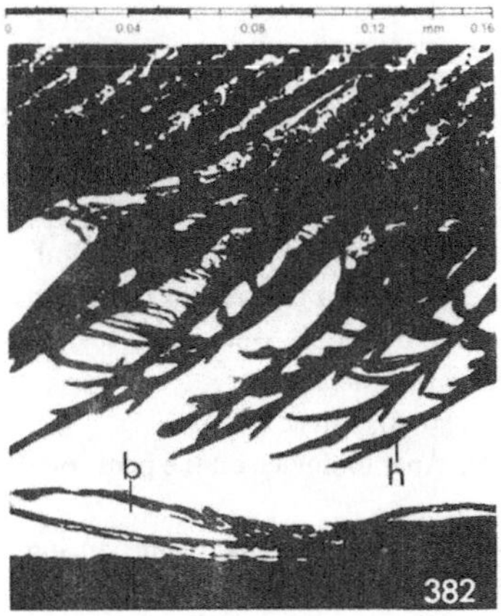

382

into a number of segments. When a bird flies through bushes the primary feathers often brush against them and are broken up in this way, but the damage is easily repaired. If the vanes of feathers were formed of an equally light continuous membrane, however, they would soon be torn by the mechanical stresses of flight and damaged irreparably.

Figs. 379 and 380 show a hook barbule and a bow barbule, respectively, at high magnification. One can see the strong ventrally recurved hooks on the former, which fit into the dorsal, longitudinally folded margin of the bow barbule as into a handhold (see the diagram of Fig. 381). Axial sliding out of the bow barbule is prevented by the fact that the "handhold" has small, inward-curving spurs at its distal end.

The modifications of this basic type of hook-and-ridge mechanism among the various orders are so different that they can be used as a systematic characteristic. To provide just one example, the feathers of pheasants are illustrated (382). The barbs are 0.08 mm wide; the bow barbules project from them at an angle of $+26°$ with a separation of 0.032 mm and the hook barbules, at $-32°$ with a separation of 0.024 mm. The ratio of the spacings of the two kinds of barbules is thus 4:3. The hook barbules each terminate in an extension about 0.12 mm long. This is equipped with 10 or more very fine hooklets (length *ca.* 0.02 mm, diameter *ca.* 0.0015 mm). The bow barbules also have free extensions in the form of transverse rods (length *ca.* 0.12 mm, diameter *ca.* 0.003 to 0.005 mm), with which the hooklets engage. This arrangement exemplifies a characteristic feature of such probabilistic attachments, at two levels. First, there is not a 1:1 relationship between bow and hook barbules, but rather a ratio of 4:3. Second, neither is there a 1:1 relationship between hooklets and rods, but a ratio of 10:1 or even more. Thus any of the hooklets can engage with any of the rods on any of the opposed barbules. It is worth noting that the sizes of the interlocking elements are in the range 10^0–10^1 μm. Engineers are now emphasizing that they will have to be concerned with structural elements of this order of magnitude.

Plate 27 (Figs. 386–405) ▶

386. Expansion anchor: the proboscis of Saccoglossus cambrensis
387. Locomotion in Saccoglossus cambrensis; p, extended proboscis; me, collar (or mesosome); m, mouth opening
388. Patent bottle stopper
389 and 390. Expansion nuts
391–395. Anchor device on the penis of Sympetrum
396. Wall plug with spring toggles during insertion
397. Wall plug with spring toggles, in position
398. Ball spacers positioned by tightening a cone-tipped screw
399. Pipe plug with spring washers
400. Multiple-jaw chuck
401. Spacer-type attachment of Freyanella plataleae between two feathers
402. Balistopus aculeatus wedged in place by the extended runner and spur
403. Flat-bed press
404. Opened stylets of Ichthyotomus
405. Wound forceps

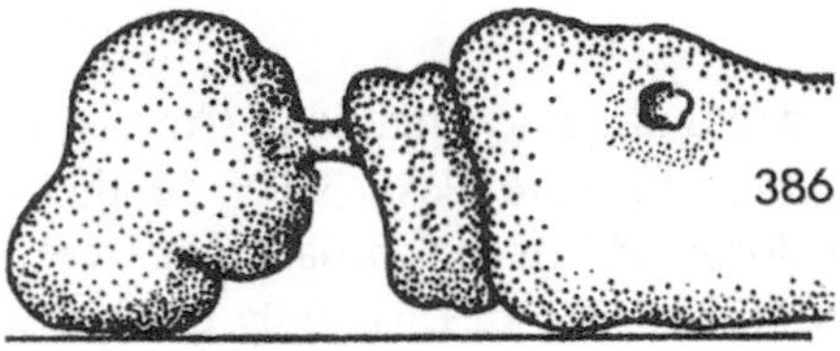
386

387

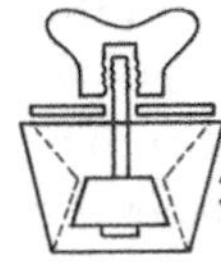
388

391

392

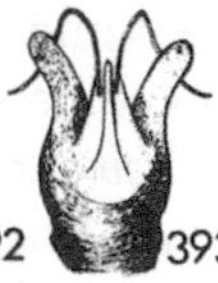
393

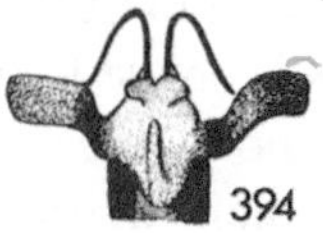
394

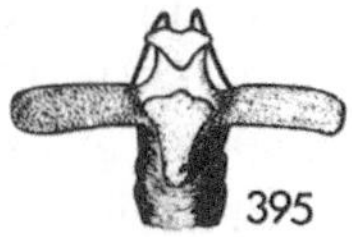
395

389

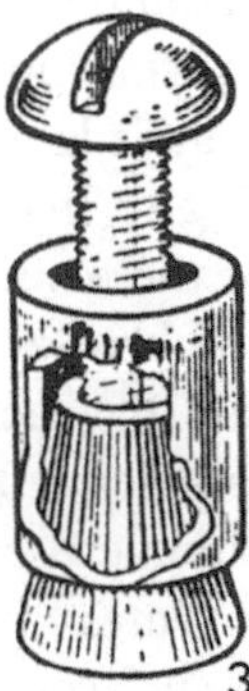
390

396

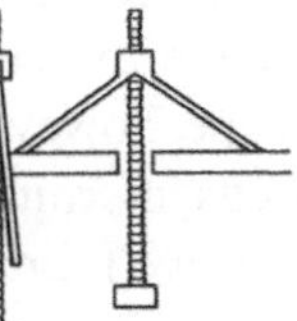
397

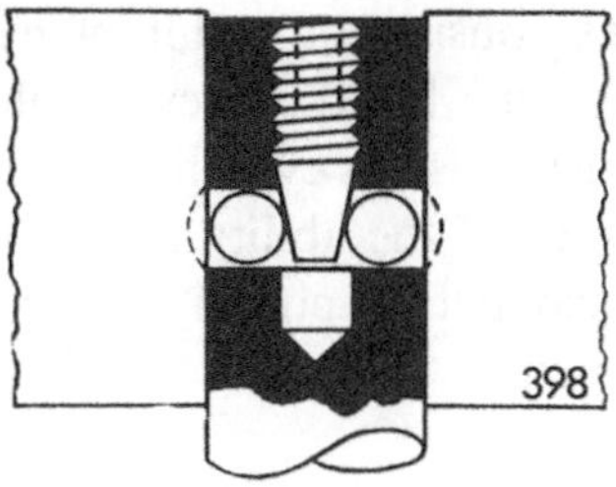
398

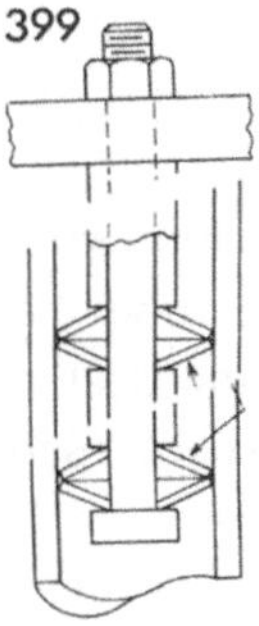
399

400

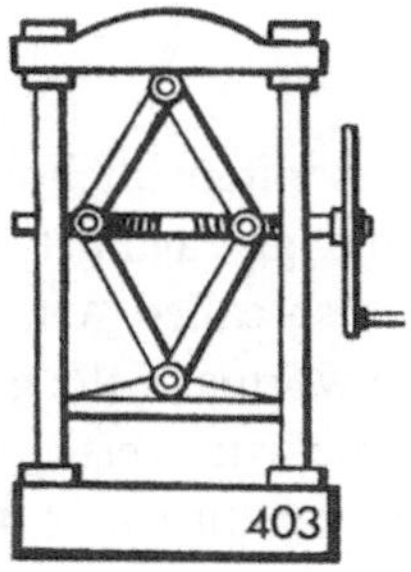
403

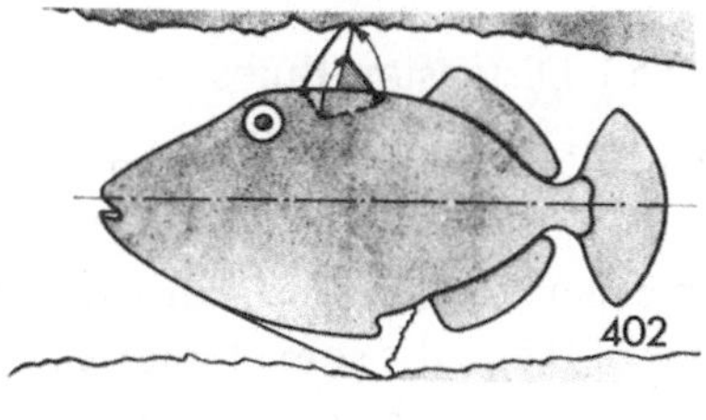
402

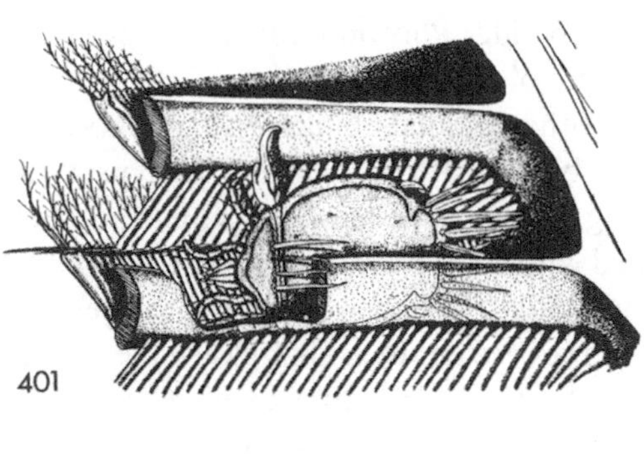
401

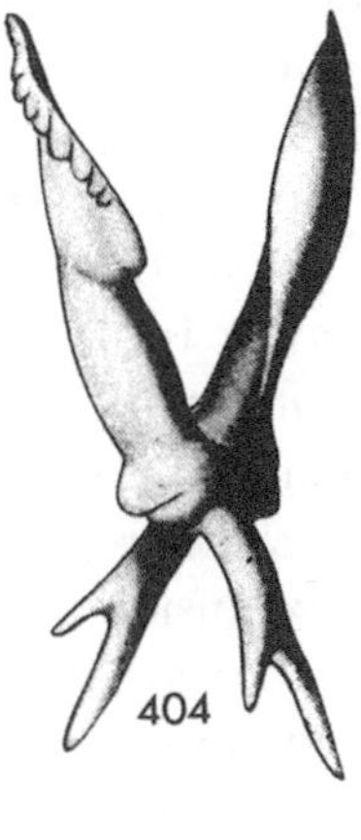
404

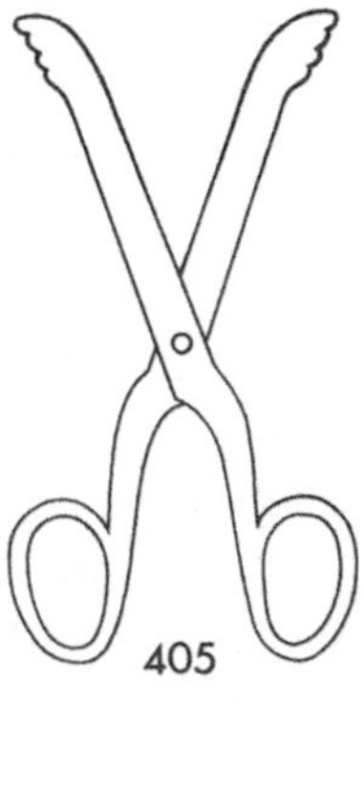
405

2.2.4.4 Microhook Principle

Among the reptiles, most of the Geckonidae and some Iguanidae (Anolis) can walk on quite smooth vertical surfaces, even on plate glass. They do this by means of attachment devices having the form of leaf-like broadenings of the fingers and toes, called digital pads or discs (383–385). Occasionally these structures are located even on the underside of the tip of the tail (genus Lygodactylus). On their surfaces, on transverse lamellae of skin, there are dense rows or "brushes" of tiny bristles (84–118 μm in length). The pads contain a system of blood lacunae. When these are filled, the brushes are pressed firmly against the substrate and become anchored in minute irregularities. On a single lamella in one species of gecko 30,000 bristles have been counted; according to that count, the whole animal has at its disposal more than two million of these very fine attachment elements. Under the light microscope the ends of the bristles appear thickened like buttons. The electron microscope, however, reveals that each of these thickenings consists of a tuft of hundreds of minute (diameter 0.05 μm!) processes bent like hooks. These could very well catch in submicroscopic irregularities which exist in even the smoothest surfaces, such as plates of mirror glass. The details of the ability of the gecko foot to adhere to surfaces have not yet been elucidated; electrical, capillary and pneumatic mechanisms are still being discussed.

2.2.5 Expansion Fasteners

2.2.5.1 Distension Bulbs

Animals which live in narrow tubes often anchor themselves by blowing up part of the body so as to press strongly against the walls of the tube. With an expansion anchor of this sort, the so-called "end bulb" (373), at the posterior end of the body, some lophophorate worms of the genus Phoronis fix themselves in the tubes they have made. These run vertically into the muddy bottom of the upper littoral zone. When danger threatens, the extended body is rapidly pulled back into the tube, and in part into the expandable end bulb itself, so that the animal is even more firmly wedged into the tube. During stormy weather or upon exposure to air at ebb tide, Phoronis can crawl even deeper into the tube with peristaltic movements of the end bulb.

This latter behavior is typical of other animals which live in tubes or on the surface of the mud; they crawl by repeatedly wedging one part of the body in place and pulling the rest along. Enteropneusta of the genus Saccoglossus deploy the proboscis, which is provided with strong muscles and can be contracted into a ball, as an expansion anchor (386). They then draw up the rest of the body, glue it to the tube wall with a secretion from the skin, release the expansion anchorage, extend the proboscis *p* far forward (387), anchor it once again, and so on. Another apt example is the worm-like animal Chaetoderma, a genus of the Aplacophora (Amphineura, Mollusca). These animals sit head-down in vertical mud tubes in the ocean floor. They also burrow in with peristaltic movements. The front end of the body acts as an expansion anchor and a ring of spines at the posterior end, as an additional anchoring device against which the worm can push.

Members of another molluscan class, the elephant-tooth shells or Scaphopoda, live on the sandy floor of the ocean. In locomotion they use the foot as an expansion anchor by digging it into the sand; it swells when distended with blood. At the same time either three lobes are spread out at its tip or a serrated terminal disc is unfolded. The foot of the clam operates in a quite similar way during crawling and burrowing; when it has been extended and its tip expanded in the mud, retractor muscles pull the body of the clam up to it. Nereis virens, a polychaete worm, digs its U-shaped burrow with the proboscis, which it jabs forcefully into the substrate as an anchor, so that during retraction the body is pulled toward it. A more common example is the expansion anchorage of the earthworm during peristaltic crawling in its tubes in the earth. This is brought about by the contraction of certain longitudinal muscles, which causes a few segments to round up; for additional security the bristles of those segments are forced into the wall of the tube.

Analogous technological devices are ampullar structures which can expand in all directions, such as the patent bottle stoppers which change their diameter by means of a screw or lever mechanism and can thus fit tightly into the necks of bottles of different inside diameters (388). Expansion nuts (389, 390) and the expanding wall plugs to be discussed below (397) also belong in this category.

The penis of dragonflies of the genus Sympetrum is composed of three parts which can be folded together; it is introduced into the vagina in this state. Then, by an increase of the internal pressure, it is unfolded so that the paired terminal structure (the so-called "glans penis") appears and a previously limp pair of "bristles" stiffens and unrolls outward. The process is illustrated in Figs. 391–395. Glans and bristles constitute a system of spreaders which anchor the penis firmly in the vagina. They function similarly, in principle, to the expansion screws and plugs which are placed in a drill-hole in the small-diameter state and then are drawn up and expanded when the screw is tightened (i.e., the "internal pressure" is increased) so as to provide a firm anchorage. This happens either by unfolding of toggles which hold firm on the back side of the drilled wall (396, 397) or by expansion of the walls of the plug itself so as to anchor the plug within the drill-hole.

2.2.5.2 Spacers

Spacers which can be pushed, folded or screwed out are sometimes used to position an object firmly between two or more other immovable objects or in the center of a ring. For example, a temporary fence can be wedged into the doorway of a child's room in this way, and it is a common method for fixing shower-curtain rods in place. Occasionally in technical devices balls are used as spacers; they are pushed outward as a central screw with a conical end is tightened and thus, for example, wedge a rod in position in a guide-hole (398). Certain commercial pipe plugs involve spring rings which, when compressed, wedge against the inside walls of pipes (399); similarly, the multiple-jaw chuck of a lathe holds an object to be turned (400). In nature this type of expansion fastener is well exemplified by the mite Freyanella plataleae. This species lives in the flight feathers of many birds. On the right side only of the propodosoma there is a broad, jointed "shoulder bristle." The mite unfolds this and presses it against a barb of the feather (401). This action pushes the body firmly against the adjacent barb, and wedges it so tightly among the barbules that it cannot be dislodged, even during the most violent flight movements.

The filefish Balistopus aculeatus (Balistidae) can extend a sort of runner on its ventral side and dorsally, a spur. These lock in position and effectively wedge the fish within fissure and between stones (402). Even when the fish is caught on a fishing line, it cannot be pulled from its haven. One is reminded of the action of the flat-bed press; in the double-screw-and-lever type shown in Fig. 403, the movable elements expand with great force against the top and bottom of the frame of the press.

2.2.5.3 Dilators

The genus Ichthyotomus is one of the few genuine parasites among the polychaete worms. As the name of the genus indicates, the animals penetrate the bodies of fishes, usually into the unpaired fins of marine eels. The worm drills into the skin with a pair of stylets pivoted like scissors. During this operation, the jaws of this chitinous drilling device are closed. The worm then expands the stylets by muscular contraction so as to hold the wound or blood vessel open. In this position the thickened pivot section, which was held apart during drilling, snaps together (404). By further contraction of the stylet muscles the "scissors" is opened until it catches in a resting position. In this way the parasite wedges itself firmly in the skin of the host. There are barbs on the tips of the stylets which further secure the connection. A number of medical instruments for surgery and examinations are designed to spread apart and hold open wounds (405) or body openings (the many varieties of specula); the principle of operation of these is of course quite comparable.

2.2.6 Attachment by Suction

Structures designed for suction attachment are usually plates or hemispheres with elastic edges which can conform closely to a smooth surface. The attachment depends upon the creation of a reduced pressure in the space between suction cup and substrate. That is, some force must be exerted which tends to increase the volume of this cavity or film of fluid. The edges of the cup must not collapse during this process; usually they have stiff margins or reinforcing rings.

One way that the requisite force may be applied is by muscular contraction arranged so as to draw the center of the cup away from the surface while the edges remain seated. The suckers of cephalopods and those on the podia of sea stars operate on this principle, as do the "plumber's friend", used for clearing stopped-up drains, and the suction mountings on small bench vises.

On the other hand, the suction-creating force may depend simply on elastic deformation of the cup itself. If a suitably compliant cup is pressed firmly for an instant onto a smooth substrate, the cup then tends to resume its concave form and in so doing generates the required partial vacuum. The suction devices designed to hold hand towels and soap on tiled walls, the suction feet on certain kitchen appliances used on smooth counters, and the cup-tipped darts fired by toy pistols all work by this principle (406–408). Biological counterparts include the prehensile apparatus of some terrestrial planarians and, probably, the suction discs on the anterior tarsi of male water beetles.

In both of the above mechanisms, work is done to reduce the pressure beneath the cup. In the first method, the work is done directly during the reduction. In the second, the external work is done in flattening the cup; the energy is then stored in its elastic deformation. When the applied force flattening the cup is removed, these elastic forces tend to return the cup to its original shape. Altogether, they act so as to oppose the external atmospheric pressure, so that if the seal at the rim is good, there results a reduced pressure (and a corresponding increase of the volume) in the material (fluid and/or air) remaining between cup and substrate. That is, an equilibrium occurs when this reduced internal pressure, plus some remaining deformation of the cup, just balance the atmospheric pressure. With a well-designed cup, the resultant distribution of forces can aid the sealing action at the rim, and the connection can be quite stable. The cup as a whole is held to the substrate by the pressure difference. Attempts to pull the cup off the substrate (as long as the seal at the rim holds) produce both a further reduction of the cavity pressure and further corresponding distortions of the cup. The connection usually breaks, however, before the pull reaches the value calculated from the atmospheric pressure and the area of the cup.

The use of suction mechanisms is remarkably common in the animal kingdom, and is found in the most diverse classes of animals. Such devices can be classified according to function as sessile or stalked single suckers, double and multiple aggregates, and finally suction plates composed of numerous individual elements. Suction cups serve most commonly for static attachment of an animal to its substrate, but other uses are notable. Occasionally they are organs of locomotion. In this case the attachment, release, and change of the spacing between two suction elements must occur in a coordinated way. Leeches, dipteran larvae, octopuses, and sea stars, for example, are capable of this behavior. Temporary suction attachments sometimes serve for the "vacuuming" up of a surface coating. The mouths of tadpoles work in this way. Finally, two systems of tubes can be coupled by a suction mechanism: a very effective suction connection between the mouth of an infant and the mother's nipple makes possible an active and reliable transfer of the milk from the ducts of the milk glands to the esophagus of the baby.

2.2.6.1 Simple Suction Cups

Among the monogenean flukes (Trematoda), for example, the Monopisthocotylinea have a large well-developed sucker at the posterior end. It is often bell-shaped and on a small stalk; there is a strong ring about the rim, as well as accessory attachment devices such as pairs of hooks (Encotyllabe, 409) or radial rows of hooks (Acanthocotyle, 410). The powerful, muscular suckers on the ventral and oral surfaces of the cercariae (411) form deep cups in the body of the animal. Large, thin-walled suckers are made more effective by mechanical reinforcing elements; a framework of curved T-shaped chitinous struts supports the sucker of a species of Diclidophora (412). In the 4-mm-long digenean fluke Anaporrhutum largum from the gut of the shark (413), the ventral sucker, which is usually at the end of the animal, has migrated centrally to a position where it can draw the slightly concave ventral side of the animal flat against the gut wall of the host. In some members of another digenean family, the Paramphistomidae, the entire ventral surface of the body forms a single giant sucker (414, 415). The cavity is surrounded by a strong ring and its surface is occasionally covered with papillae. These nearly hemispherical animals live in the gut of the horse and the hippopotamus, where this gigantic sucker serves as a very efficient organ of attachment.

The suction organs of the cestodes (tapeworms) are well known; they appear in paired, fourfold, or multiple arrays. Some Tetraphyllidea have, in addition to their four lateral suction organs, a single central sucker on the scolex which

Plate 28 (Figs. 406–424) ▶

406–408. Toy arrow with suction tip
409. Hooks in the sucker of Encotyllabe
410. Rows of hooks in the sucker of Acanthocotyle
411. Oral and ventral suckers of the cercaria of Dicrocoelium lanceolatum
412. Sucker of Diclidophora spec.
413. Ventral sucker of Anaporrhutum largum
414. Giant sucker of a genus of the Paramphistomidae, side view
415. Giant sucker of (414), ventral view
416. Scolex of Lecanicephalum abruptum (Tetraphyllidea); apical sucker and two accessory suckers, everted
417. Scolex of (416) with suckers drawn in
418. Plunger
419. Suction foot for a bench vise
420. Sucker of Pterodectes
421. Suction-cup towel hook
422. Child's arrow
423. Foot of a kitchen appliance
424. Suction plate for a windshield attachment

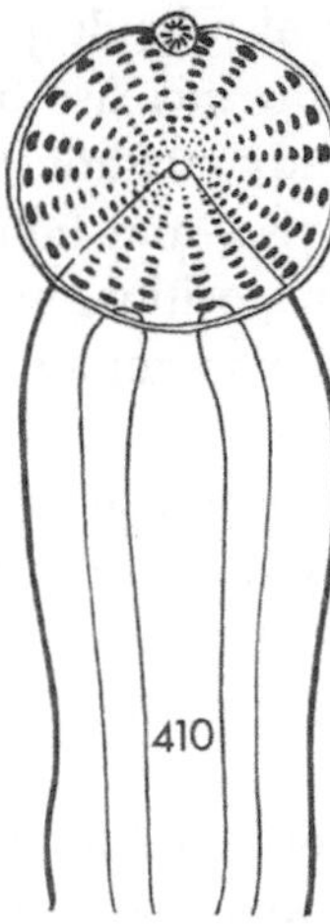
410

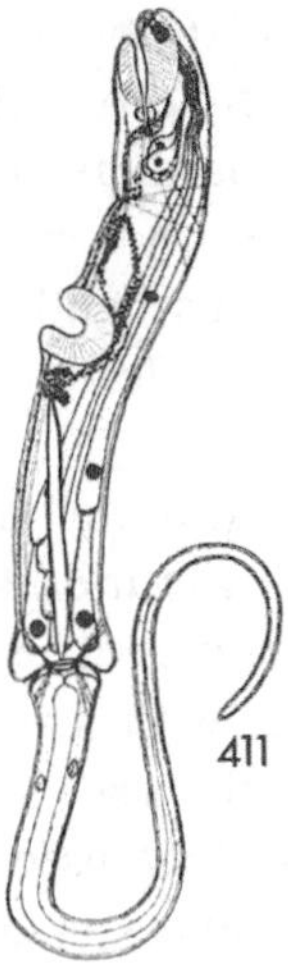
411

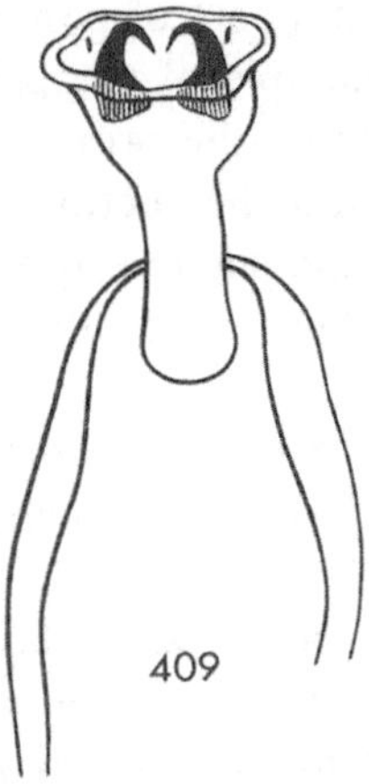
409

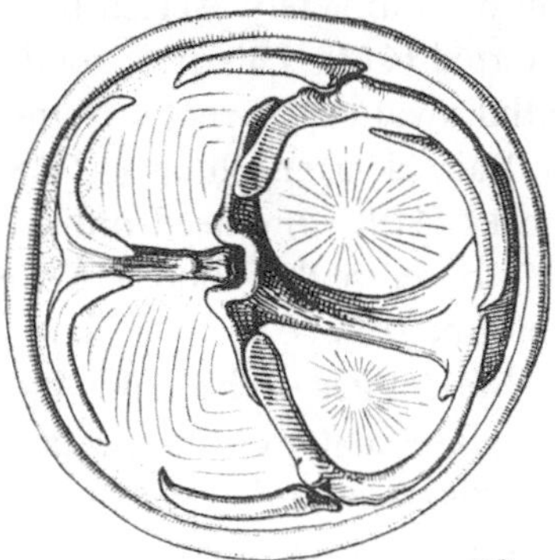
412

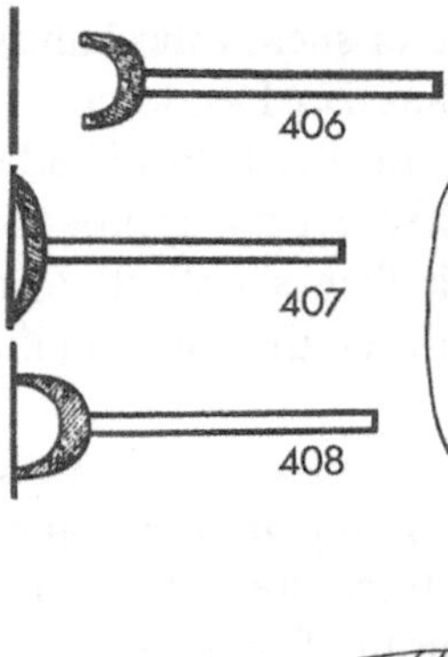
406

407

408

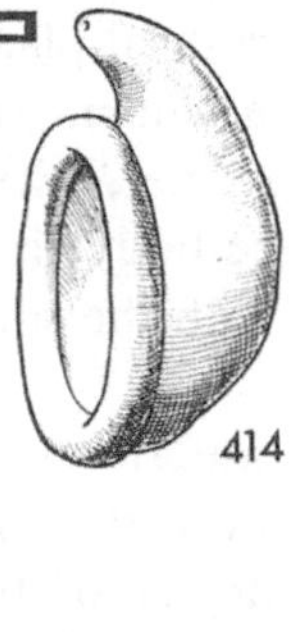
414

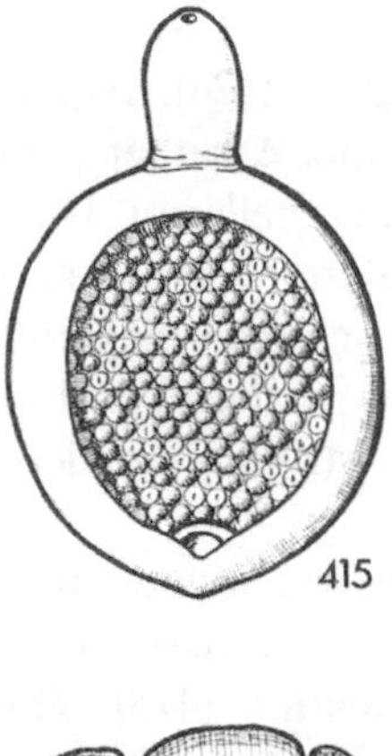
415

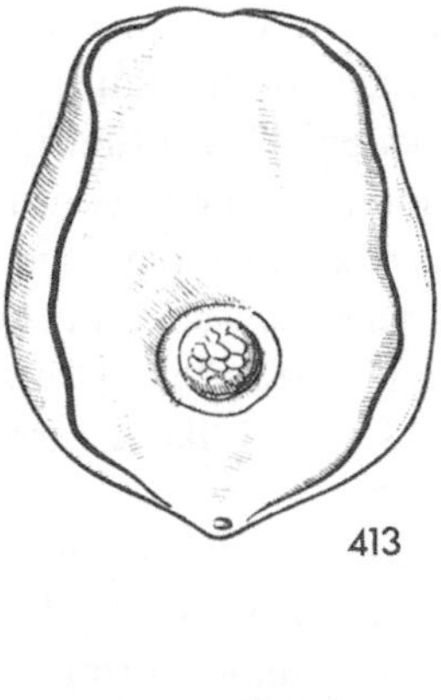
413

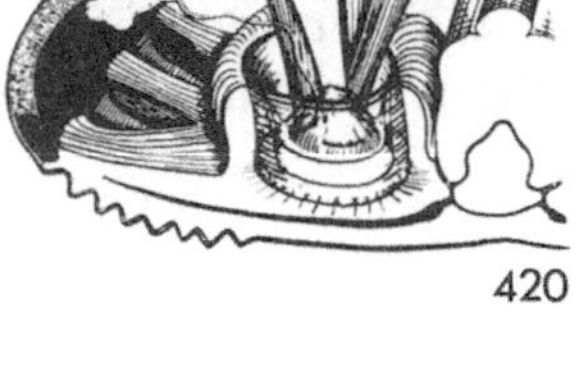
420

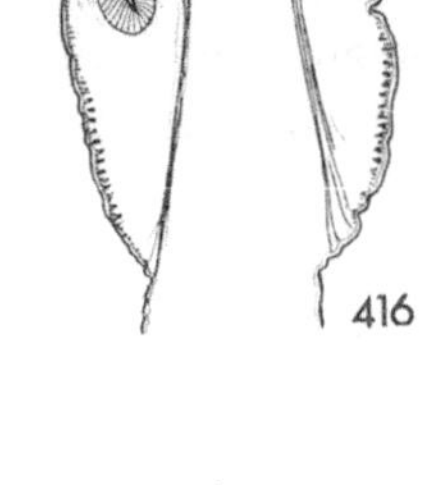
416

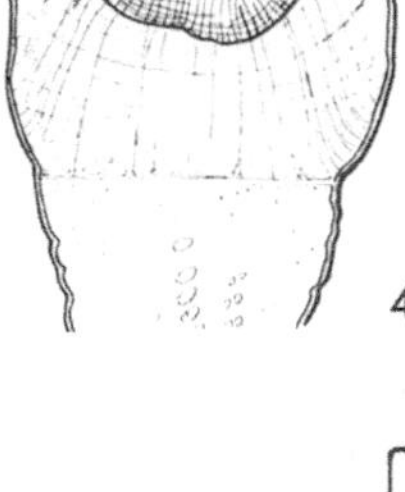
417

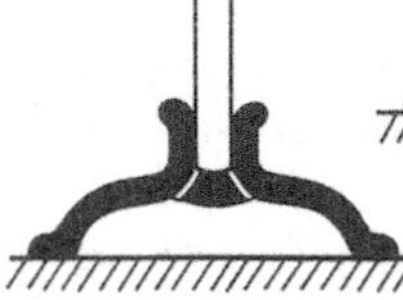
418

419

422

421

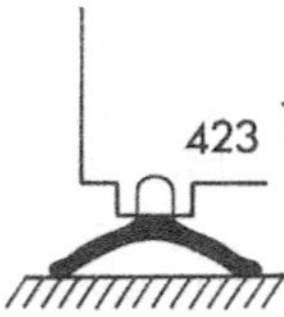
423

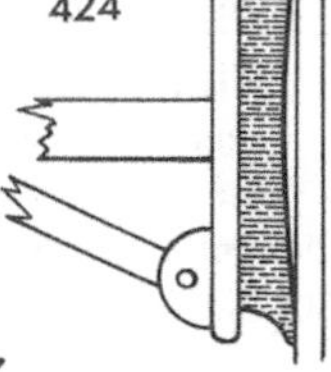
424

can be completely everted to form a sort of pestle (416) but can also be deeply inverted (417). This is a good example of the principle of volume increase by withdrawal of the center of the cup—the principle used, as already mentioned, in drain-clearing plungers (418) and the suction feet on bench vises and filing blocks (419). One need only set the suction apparatus lightly in position, and then by muscular contraction or lever action draw the flexible diaphragm away from the surface.

Male mites of the genus Pterodectes have remarkably well-developed suckers (420). The center of the sucker surface can be raised by four dorsoventral muscles and displaced sideways by three lateral muscle bundles. The wall around the sucker is also well equipped with muscles. A gland, with its duct opening into the cavity of the sucker, secretes a fluid to improve the seal.

The second suction principle mentioned above, that used for towel-hooks (421), children's arrows (422), feet for kitchen appliances (423), and plates for attachment to panes of glass (424), requires that the cup be firmly pressed or clapped onto the substrate. A rather peculiar example of such a mechanism is found in the flatworms (Turbellaria). The terrestrial planarian Dendrocoelum has a small pit on the anterior ventral surface which, with the help of radial muscles, can act as a sucker. According to Kaestner, when a prey approaches the planarian draws up the posterior end of the body; it then shoots its front end forward like an arrow, striking the victim with the sucker and thus holding it firm.

Suckers of various sizes are found on the anterior tarsi of the male water beetle Dytiscus. The beetle uses these to lock onto the smooth thoracic shield of the female during copulation (425). The largest of these suckers is on a stalk on the first tarsal segment. In the course of chasing the female, the male strikes her violently with them. Like all suction devices of insects, these giant suckers (diameter *ca.* 1.5 mm) are purely cuticular structures (426). A huge epidermal cell forms an axial cylinder which is buried deep in the tarsal segment and consists of 12 brown-pigmented chitinous rods; these surround the soft, almost colorless axial chitin and radiate out into the soft chitin of the disc

Plate 29 (Figs. 425–433) ▶

425. Copulatory position of Dytiscus
426. Giant sucker of Dytiscus
427. Giant sucker of Dytiscus marginalis with reinforcing struts, seen from below
428. Small sucker of Dytiscus
429. Anterior tarsus of Dytiscus
430. Elements of the suction disc in the large sucker of Dytiscus marginalis; cross section through the stalk
431. Radial vertical section through the edge of the disc (from 430)
432. Diagram of the reinforcing structures of the disc in (430); only four of the ribs are shown complete with all their branches
433. Suction surface of an oval miniature sucker on the anterior tarsus of a male Dytiscus marginalis; diameter *ca.* 0.2 mm

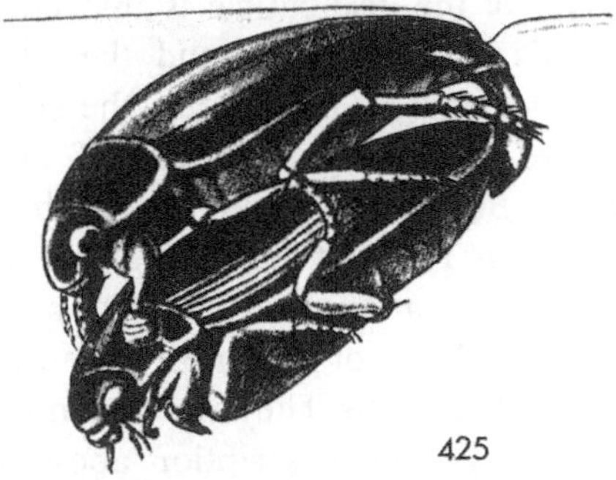

425

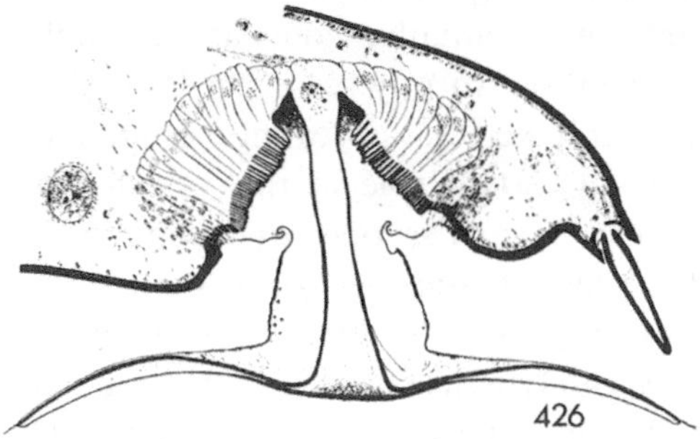

426

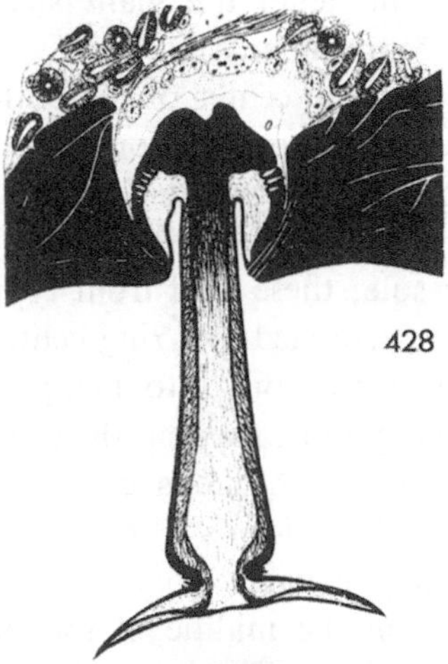

428

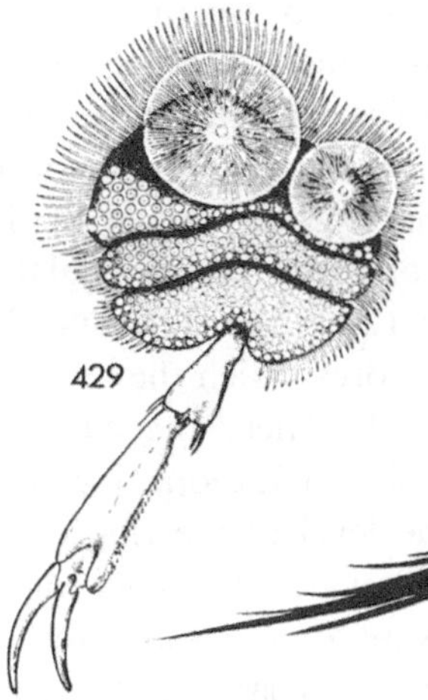

429

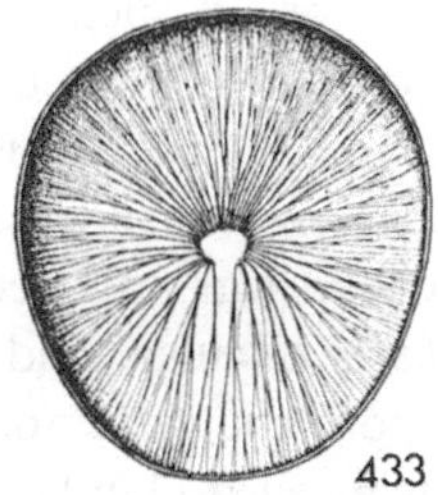

433

431

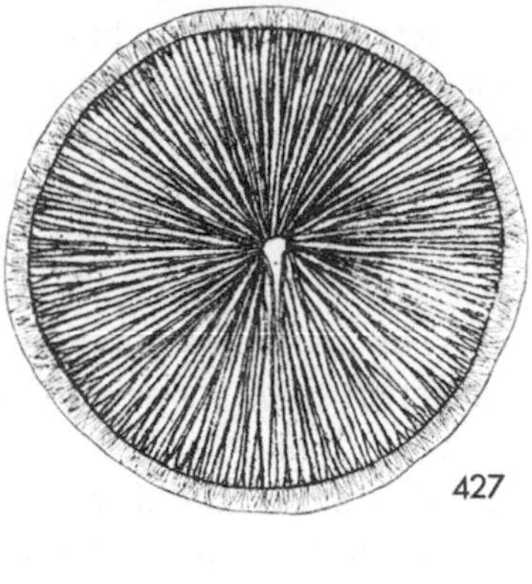

427

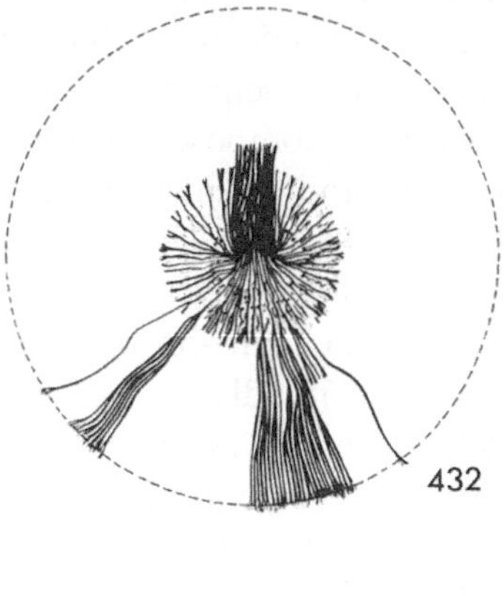

432

430

as reinforcing struts (427). The thin outer layer of the disc cuticle is thickened only in an annular zone at the stalk. It ends in a fringe around the edge of the disc. An oily secretion from single-celled glands at the base of the stalk probably acts as a "sealing compound." As we shall see below, "gluing" also plays a certain role in this attachment organ. Beside this giant sucker there is also a smaller one (diameter *ca.* 0.75 mm) on the first tarsal segment of the male Dytiscus. Although on a smaller scale, its construction is functionally similar (428). The best investigated suckers are those of the most common species, Dytiscus marginalis (429), and of Dytiscus latissimus. They are fundamentally the same in these two species. Though the general description above is valid for all such suckers, it is worth describing in detail the giant sucker, probably the most conspicuous attachment organ found in any insect.

The sucker consists of a dish-shaped suction disc, the center of the bottom of which protrudes somewhat from the basic concave cross section; it is on a cylindrical stalk set deep into the sole of the first tarsal segment. The suction disc is circular, and is oriented perpendicular to the stalk. A large number of pore channels open between the stalk and the sole; these lead from typical single-celled glands. The pores which they serve are arranged in a ring centered upon the axis of the stalk. There the cuticle is drawn out into many tiny rigid bristles; these may shelter the openings of the glands, keeping them from being stopped up. At the level of the ring of pores the chitinous covering of the sole splits into two layers. The outer one of these forms a ring-shaped fold, continues as the wall of the stalk and finally expands to become the outer wall of the sucker. This layer is thin; only in the middle of the stalk is it noticeably thickened by a broad ring of lighter chitin. The inner chitinous layer makes a bell-shaped arch deep into the tarsus where it forms the abutment for the stalk within the tarsal depression. This inner layer may be termed the "palisade layer", since it is composed of numerous radially arranged tubules which are closed distally and open proximally. These tubules are formed by the columnar epithelial cells, with large nuclei, which lie above them. In the center of the invaginated inner layer there arises a concentric bundle of, on the average, 12 strong, elastic, dark-brown chitinous rods. In cross section these appear as a ring near the middle of the stalk, centered upon its axis (430); together they are called the axis cylinder. Numerous anastomoses are present. Up to the floor of the disc, the chitinous rods run in parallel; they then bend outward and radiate (432) into the vicinity of the margin. But they end short of the edge; the delicate marginal fringe of the disc is a separate structure (431).

These chitinous rods form a strong inner skeleton of reinforcing ribs. They run along the outer surface of the disc, toward which they send fine branched supporting fibers, and there continue to branch like a tree (432); this arrangement is also found in the smaller suckers (433). The axis cylinder and radial ribs are embedded in a matrix of soft, almost colorless chitin, which fills up the stalk and the entire volume of the disc. The axis cylinder is actually formed by a single giant cell. In the pupal stage this cell extends to the floor of the disc; it then withdraws progressively, secreting both the chitinous supporting elements and the soft matrix. Finally it deposits a plug and then a cap of chitin on "the cupola formed by the palisade layer, like the keystone of an arch"—as H. Blunck has so graphically described it. Then the cell disintegrates. Stalk and cup are entirely chitinous structures. In use, the cups are slapped flat onto the female, a stone, or the wall of the aquarium. This bends the chitinous radial ribs of the disc. Restoring forces tend to return the disc to the original dish-like shape. This enlarges the central cavity somewhat, and the reduced pressure necessary for attachment is created. The secretion which emerges from the ring-shaped pore zone is an oil of low viscosity which, because of its tendency to spread out, covers the entire apparatus. It acts as a sealing compound and at the same time as a supplementary glue. The attachment apparatus of Dytiscus is thus a combined suction-and-glue system, comparable to the podia of echinoderms. This has been demonstrated experimentally by de-greasing with ether and by tests in a depressurized chamber. There is a further discussion of the holding capacity of the disc on page 152. The biological significance of the suckers as an aid to copulation has also been demonstrated; males from which the large suckers have been removed cannot hold onto the smooth back of the female. Ordinarily, the pairs remain coupled for 24 hours or longer; the suction attachment, once made, must hold for at least this long. The marginal fringe and the layer of oil, which help to compensate for small surface irregularities, appear essential to this kind of performance. In fact, it is doubtful that any such suction apparatus in the animal kingdom operates entirely without supplementary secretion.

To release the suction, the male slides the still firmly-attached tarsi to the side, away from the attachment position on the thorax of the female, and simultaneously makes powerful sideways jerks of the body. It has also been observed that the suckers can play a role in the capture of small fish.

Most leeches have large "adhesive discs" (434) on the posterior end, with which they can remain attached for considerable periods. (Note that the words "adhesive" and "adhere" are used by common convention in the description of connections which need not depend entirely upon the physical forces of adhesion; see, e.g., Section 2.2.8) They also use them for inchworm-like "walking." The discs almost always represent seven fused segments, though in one genus there are only four. Only in Lumbricobdella are they lacking. The musculature consists of concentrically arranged "equatorial" muscles which constrict the cup, "meridional" muscles which lift its floor, and "radial" muscles which are stretched between the outer and inner walls of the disc (435). When the radial muscles contract before the cup is placed on a surface, they flatten the suction apparatus and thus enlarge it. The way the disc is placed on the substrate and the presence of a viscous mucoid glandular secretion suggest that attachment by suction is not the only principle involved. Indeed, this notion is supported by the observation that some species of leech do not fall from the substrate when the pressure in a chamber is lowered.

Suckers for mutual positioning of copulating partners appear not only in mites, water beetles, and cephalopods, but also in worms. In the spiny-headed worms (Acanthocephala) the male copulatory organ lies in a deep pocket at the end of the body. The pocket can be everted by muscular contraction so as to form a hollow goblet-like or bell-shaped organ. The male uses this device to attach himself to the posterior end of the female by suction and thus positions the two genital openings opposite one another.

The sucker-like foot of the common river limpet Ancylus fluviatilis should be mentioned as a special case. This gastropod lives on the stony bottom of fast-flowing mountain streams. Without its suction apparatus it would be washed away. Most chitons (Polyplacophora; 436) live in the surf zone of the oceans. They apply the flat ventral surface, from mantle groove *g* to foot *f*, to the rocks. By contraction of the dorsoventral muscles *dm* they create suction spaces between foot and substrate and also enlarge the gill chamber *gc*. In addition to these suction mechanisms, glue and direct intermolecular adhesion play a supplementary role. This is doubtless also true of the curious "honeytail", Dulcicauda griseaurella (Rhinocolumnidae, Monorhina). These animals (438) attach themselves by means of the broadened nose to a pillar-like pedestal called the sella, which is secreted by special skin glands within the suction surface. They lead a sedentary life and break the attachment only during the mating season. Unfortunately, souvenir-hunting humans have decimated these singular inhabitants of tropical islands, so that closer investigation of their attachment mechanism appears unlikely.

Dassies (small African mammals of the genus Hyrax) can flit along smooth rocks as quickly as the eye can follow; they appear to run effortlessly up vertical and even overhanging walls of rock. The blackish soles of the feet of these animals are of a rubbery elasticity and have several callous pads separated by deep fissures. The central fissure can be contracted or expanded, so that the animal can exert suction upon a smooth substrate and then release it. The skin of the soles is extraordinarily well supplied with sweat glands. Per unit surface area there are 15 times as many glands as on the sole of the human foot. This watery secretion provides the fluid film which is always to be found in association with the suction devices of land animals. These attachment organs, unique among mammals, long ago excited the interest of students of natural history. G. Schweinfurth wrote in the description of his travels, which appeared in 1874, "If one has shot [a dassie] and goes to pick it up, it clings with its feet to the smooth rock in a dying battle as though it had grown there."

Among the mammals, the apes and lemuroids also make use of suction surfaces, though their degree of specialization does not approach that of the

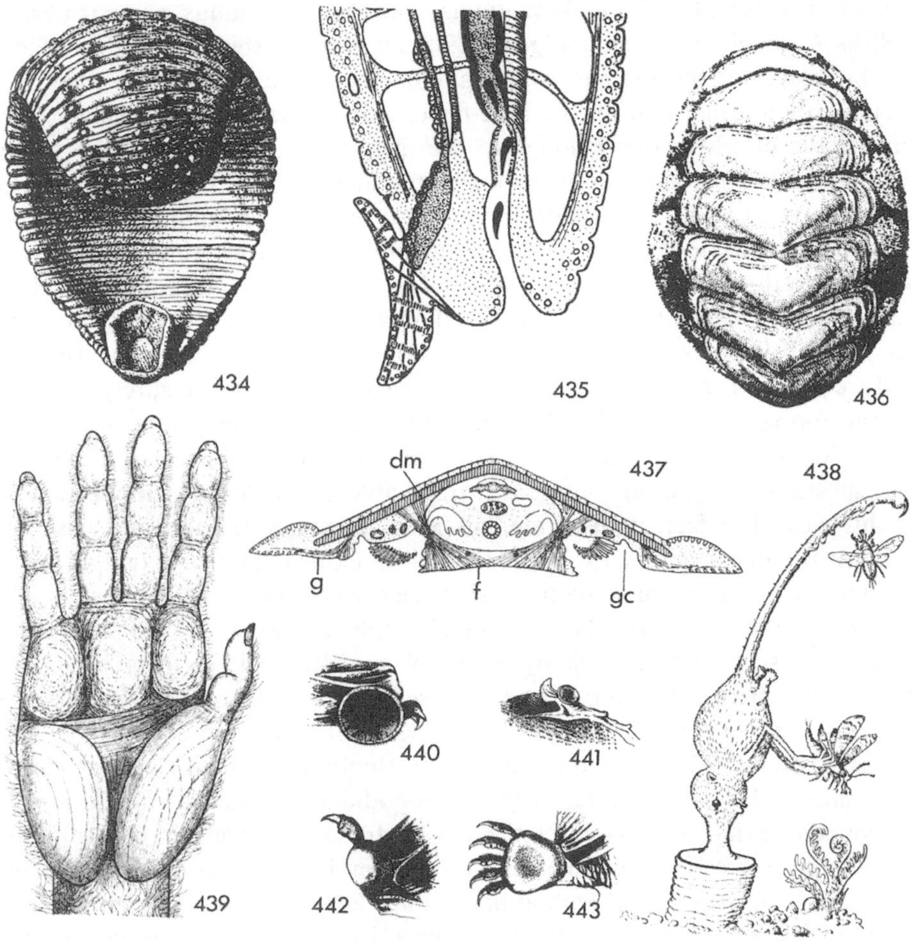

Plate 30 (Figs. 434–443)

434. Hemiclepsis marginata
435. Adhesive disc of a leech
436. Acanthopleura spiniger (Polyplacophora)
437. Semidiagrammatic cross section through Lepidochiton marginatus
438. Adhesive nose of Dulcicauda griseaurella
439. Adhesive pads of Cercopithecus cynomolgus
440. Suction disc on the first finger of Thyroptera tricolor
441. Pedal disc on the metatarsus of Thyroptera tricolor
442. Disc on the first finger of Tylonycteris pachypus
443. Pedal disc of Tylonycteris pachypus

dassies. The long-tailed monkeys Inuus speciosus and Inuus nemestrinus, as well as Cercopithecus cynomolgus (439), can run up steep, smooth surfaces. On the inside of the forehands and hindhands of the first species there is a thumb ball, another ball on the little-finger side of the palm, and finally three small balls of which one belongs to the little finger, the middle one to the two middle fingers, and the outer one to the index finger. The two last-named species have in addition pads on the inside of each finger joint. Cercopithecus never embraces smooth palm trees with its limbs, but rather simply presses the hands onto the trunk. It cannot dig its claws in, since—as in man—the claws hardly extend beyond the ends of the fingers. So one can only assume that the hands adhere by the principle of suction. The pads are soft and rubbery, and are pressed firmly onto the substrate. The reduced pressure is surely produced by the formation of cup- or dish-like depressions on the palms of the four hands. Moreover, sweat glands are found here in great numbers. The Lemuridae and species of Tupaia are also said to be able to move on smooth surfaces in this way. The feet of these animals show comparable folds and cushions. Even in the flying mammals, the bats, well-developed simple suction cups can be found, particularly in some tropical species which hang within dense tangles of smooth leaves. The Brazilian species Thyroptera tricolor, when at rest, hangs from steep, smooth rock walls by means of a sucker. All the members of this genus (two species, the above and T. discifera) have round suction discs at the wrist and on the sole of the foot (440–443). They are located on little pedestals and are used by the animals in climbing. Unstalked suction discs are found in the only member of the Myzopodidae of Madagascar, the species Myzopoda aurita. The cover of the ear in these bats ends in a small disc. The existence of suction discs in some vespertilionid bats has also been suggested, for example in the Indian bamboo bat Tylonycteris pachypus.

Suction attachments also occur in the micro-world. Some representatives of the peritrich ciliates have suction devices which are enormous in comparison with the length of the animal. Trichodina pediculus, the "polyp louse", is one of the few motile peritrichs. Often several of them live on one of the fresh-water polyps Chlorohydra viridissima (444). They glide about over the body surface like hovercrafts on their posterior circlets of cilia, while rotating slowly. In so doing, they cannot avoid touching the cnidocils which trigger the dangerous nematocysts of the host, but for reasons as yet unknown they do not discharge these "land mines"—which otherwise respond to the most delicate of touches. The Trichodina feed either on the mucus secreted by Chlorohydra or on the bacteria and fungi within it. The underside (actually the flattened posterior end) of the animal has become an attachment surface of circular outline (445). The center of the circle is slightly recessed and includes the so-called attachment ring, with an inner and outer circle of teeth. The periphery is built up in steps and is composed of an inner and outer membrane (*im* and *am*, respectively, in 446), the inner and outer vela *iv* and *av*, and a characteristically striated

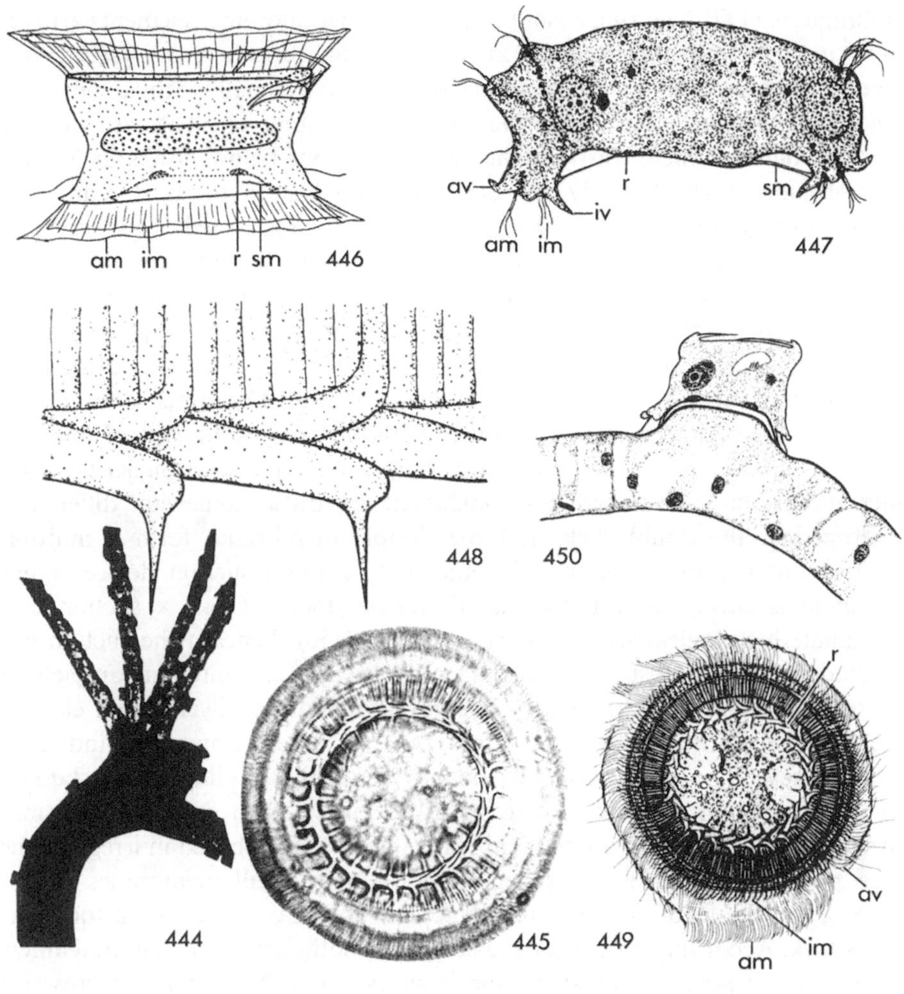

Plate 31 (Figs. 444–450)

444. Individuals of Trichodina pediculus on Chlorohydra viridissima
445. Adhesive surface of Trichodina pediculus, with ring and circlet of cilia
446. Side view of Trichodina pediculus
447. Sagittal section of Trichodina pediculus
448. Part of the ring structure of (447), seen from below
449. Disc surface of Trichodina pediculus, from below
450. Section through the "foot" of Trichodina on the host

membrane *sm* (447). The ring *r* (448) consists of inner and outer elements arrayed loosely in rows with the elements at angles to one another; considered as a whole they form a circular structure which is anchored by small toothlike processes in the ectoplasm on either side. The bases of the inner elements are inserted under the outer elements so that the ring, seen from below, looks as though it were braided (449). Distally, the elements each bear a tooth, 4–5 μm long and with a fine point at the tip. The ring is evidently constructed of two different substances, a pellicle on the outside and protoplasm within. Around all the components of the disc surface lies a narrow, closed band of the striated membrane. The inner and outer vela are also closed rings, of flexible protoplasm, which are probably meant to seal tightly to the substrate. Two other undulating ring-shaped membranes are found between the two vela. The outer one extends far beyond the edge of the body. Both membranes are composed of uniform, fused rows of cilia, since they arise from simple rings of basal nuclei (Wetzel). Other authors have described this attachment apparatus somewhat differently.

Altogether, the double vela and the double membranes form a multiply secure seal at the rim, a feature demanded by a good suction device. When the animal is sitting at rest, the disc doubtless does act like a suction cup, as fortunate histological sections have indicated (450); beneath the suction cup the ectoderm of the host has a distinct outward bulge. Quite similar pictures exist of Trichodinopsis paradoxa on the gut epithelium of Cyclostoma elegans. Here too, the epithelial cells bulge into the attachment apparatus and fill it up completely. The mechanism of the hovercraft-like gliding is not known. Nor is the function of the odd ring structure known with certainty; one may well suppose that when the animal is at rest its teeth are anchored in irregularities of the polyp's surface or—in other species—on the gill membranes of fish or clams. But the teeth are pellicular structures and may therefore be too weak for this task. According to one of the several hypotheses about the attachment mechanism, the posterior ciliated zone is supposed to beat so as to press the animal against the substrate. It is thought that this pressure can be opposed by antagonistic inward beating of the striated ring membrane so as to create a surge of water which lifts the animal away from the substrate again. None of these attempts at explanation is really satisfactory.

Trichodina belongs to the peritrich family Urceolaridae, which includes other genera with posterior discs. Cyclochaeta and especially the type-genus Urceolaria offer good examples. The species Urceolaria korschelti is illustrated in Fig. 451 as viewed from the side; the prominent development of the lower attachment apparatus is striking. In Fig. 452 it is shown from above and in Fig. 453, from below. A system of radial ribs *rr* dominates the attachment rosette. The ring *r*, consisting of 6–22 elements, lies in a circular furrow in the rosette. Each of the sickle-shaped elements of the ring is arranged so that its concave inner surface partially overlaps the convex outer surface of the adjacent element. This ring does not bear teeth. One velum and one aboral circlet of cilia *ac*

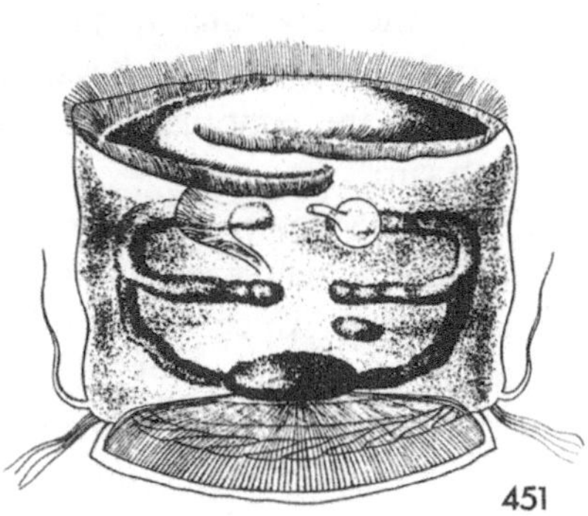

451

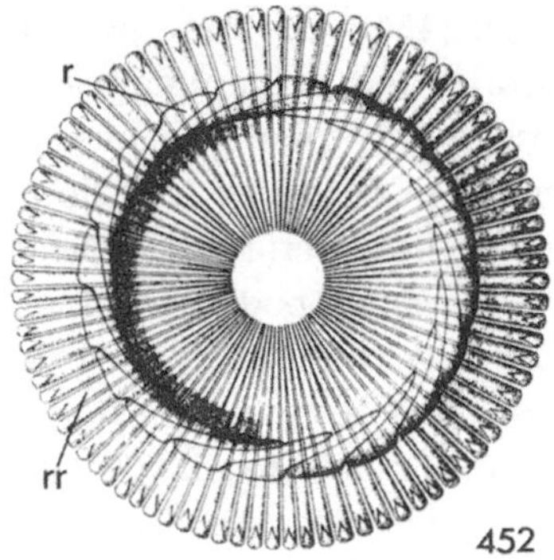

452

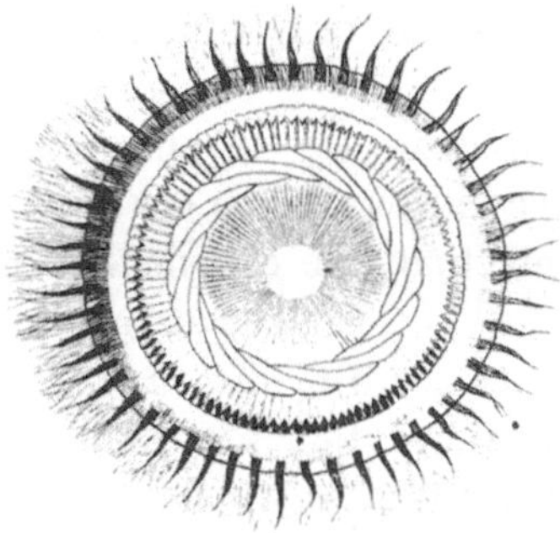

453

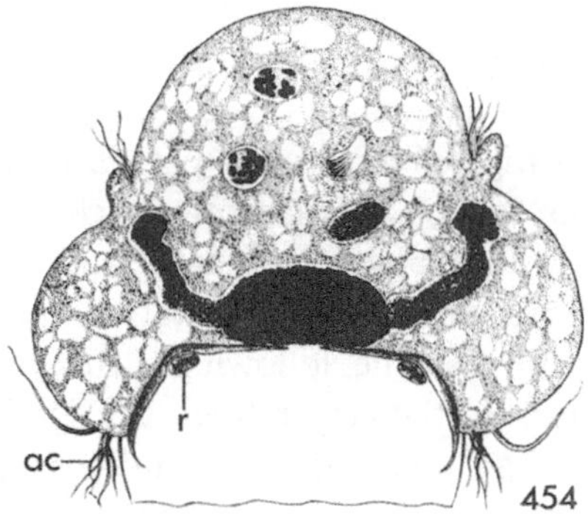

454

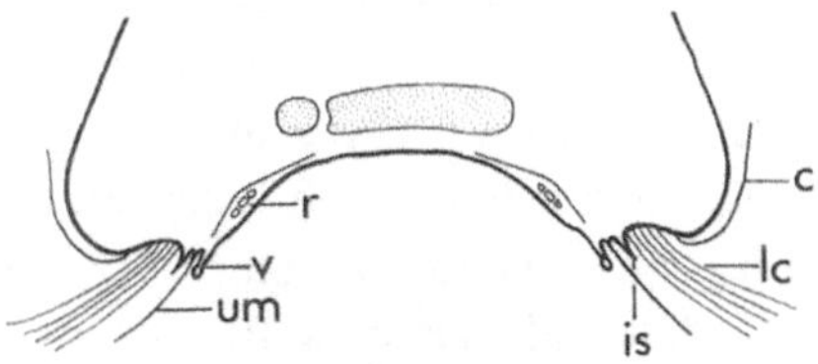

457

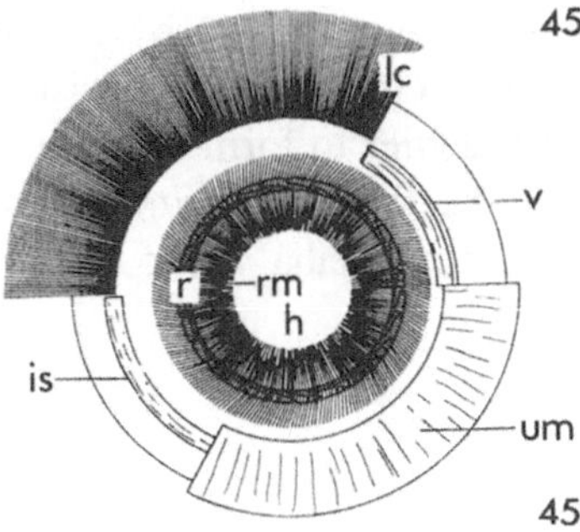

456

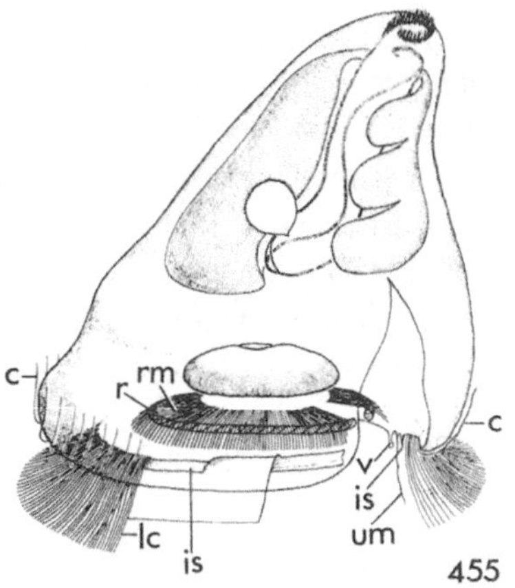

455

Plate 32 (Figs. 451–457)

451. Urceolaria korschelti
452. Posterior disc of Urceolaria korschelti, from above
453. Posterior disc of Urceolaria korschelti, from below
454. Longitudinal section through (451)
455. Urceolaria paradoxa
456. Posterior disc of Urceolaria paradoxa; schematic view of the underside
457. Posterior disc of Urceolaria paradoxa; schematic longitudinal section

are present (454) and these provide a tight seal to the substrate. It has been demonstrated that the animal can attach itself firmly by suction; the required low pressure is probably achieved by contraction of the endoplasm and the myonemes which lie over the "floor" of the suction cup.

The ring itself certainly serves as a reinforcement. A comparable structure is found in cephalopod suckers, such as those of the squid Sepia; these have notched horny rings on the margin of the inner surface which are strikingly similar in cross section. The margin of the attachment rosette of Urceolaria, like that of Trichodina, is provided with strong teeth. In both animals the marginal hooklets may press into the surface of host or prey and offer a supplementary anchorage. The structure of the ring elements is compatible with the idea that the convex-concave sliding surfaces allow internal readjustments which improve the fit to irregularities in the substrate.

The attachment disc of the species Urceolaria paradoxa (455–457) has recently been investigated more closely (Beinbrech). This species lives in the gut of snails, where its suction mechanism attaches it so firmly that one can neither suck it away with a pipette nor loosen it with a needle. The parasite damages the host tissue, and a heavy infestation can lead to the death of the snail. In the attachment disc the following structures are found (cf. Fig. 456, starting at the center): in the center there is a "hole" *h*; that is, a circle which is without further detail under the light microscope. Adjacent to this is a striated ring membrane *rm* made up of narrow connected ridges, 12.8–15.7 μm long and somewhat convex as viewed from the outside. Altogether, there are 220–280 such ridges. Next follows the typical hooked or toothed ring *r* which is sometimes also called a "corona." It is 2.6 μm thick and has an outside diameter of 50–60 μm. There are 34–36 teeth in the row. The individual elements lie one over the other like tiles on a roof. The suction cup is surrounded by a border, the velum *v*, and outside this by three concentric circles each consisting of one row of cilia. The inner circle of cilia is glued together to form an undulating membrane *um*. The next circle, separated from the first by a fine interciliar septum *is*, consists of strong, free-standing locomotor cilia *lc* up to 25 μm

Plate 33 (Figs. 458–468) ▶

458. Suction attachment in Anoplophrya paranaides
459. Ventral surface of Hoplidophrya falcifera
460. Sucker of Haptophrya gigantea
461 and 462. Suction disc of Licnophora macfarlandi
463. Loranthus; seedling with adpressorium and cladogenous root growing along the branch
464. Ruffled suction apparatus in a cestode
465. Barrel-shaped suction apparatus in a cestode
466. Glands (black) in the double sucker of a cestode
467. Argulus foliaceus
468. Suction organ of Argulus foliaceus

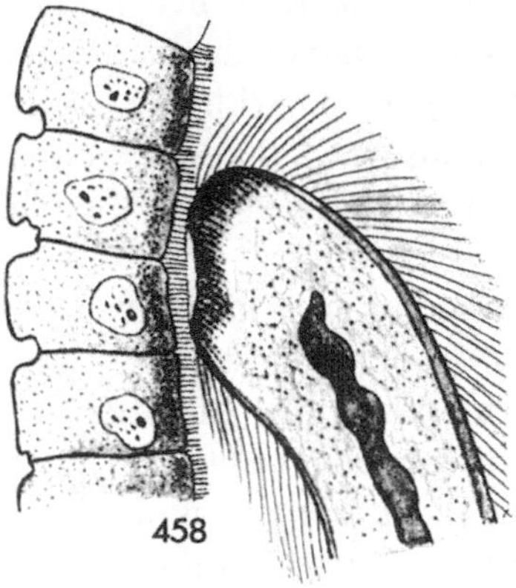
458

460

463

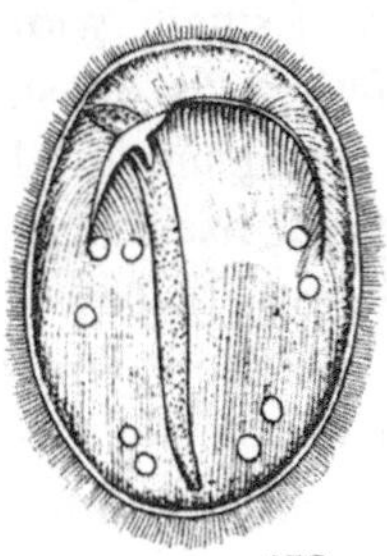
459

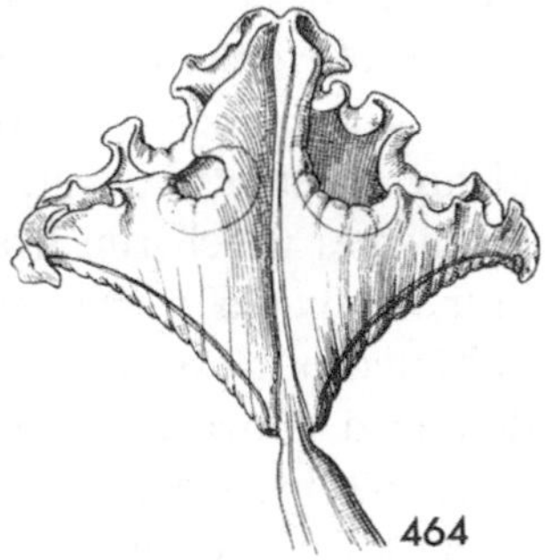
464

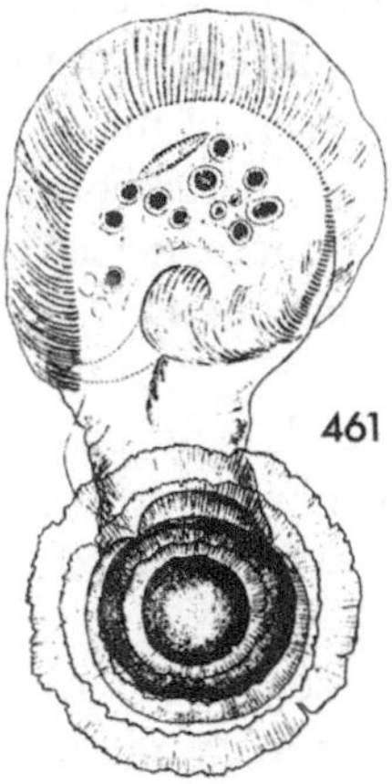
461

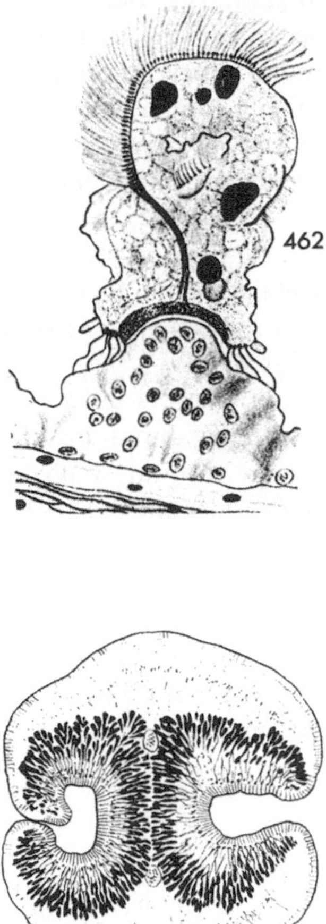
462

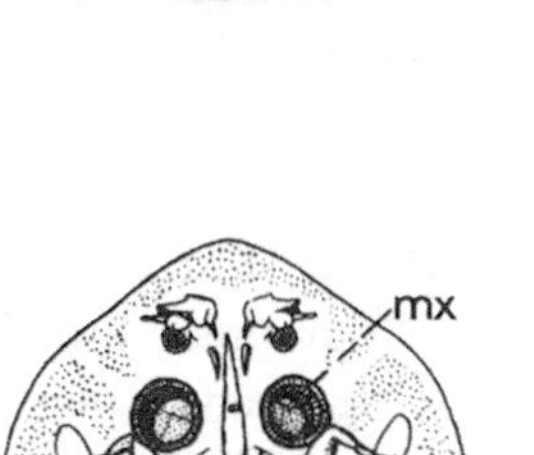
468

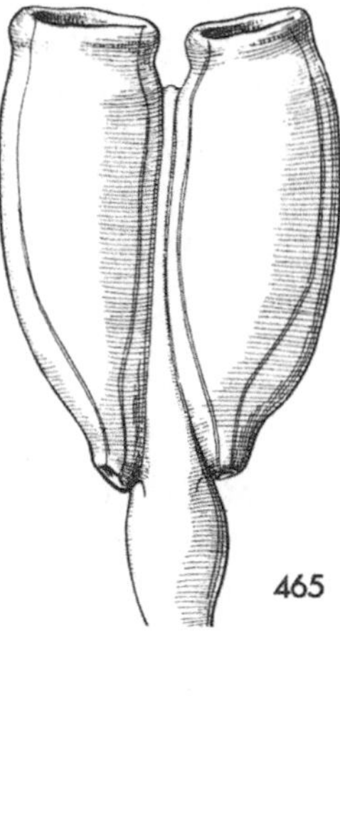
465

466

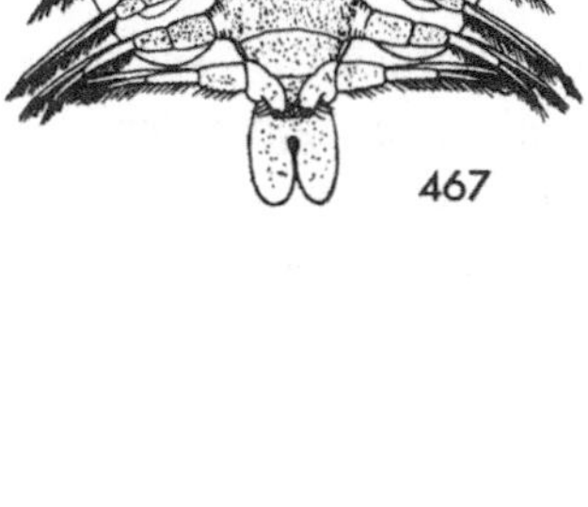

467

long. Directly above this is the insertion of the third circle, which is made up of single, widely separated, so-called "tactile cilia" *c*; although the function of these is questionable, they certainly have nothing to do with the suction mechanism. In this species, too, it has been demonstrated that the animals fix themselves in place by suction, whereas the mechanics of their "surface gliding" locomotion are unknown.

Suction cups of more simple construction are found even in the flagellate protozoans; an example is the ventral suction pit of Lamblia intestinalis. Among the ciliates, Anoplophrya paranaides exhibits possibly the earliest instance of attachment by suction. After it has stuck onto the mucous membranes in the gut of the host (an oligochaete worm) the point of attachment is drawn in, and becomes somewhat concave (458). The trench-like depression on the ventral surface of Hoplidophrya falcifera (459) is reminiscent of the situation in Lamblia. Other gut-dwellers such as Steinella, Discophrya, Haptophrya, and Lada employ highly differentiated suction organs. Haptophrya gigantea has a number of myonemes which run from the wall of the suction cup to the opposite edge of the body (460); these correspond closely to the radial muscles of the suckers of Metazoa. Nevertheless, the highly complex suction cups of the urceolarioids described above represent the peak of development. Licnophora macfarlandi (461, 462) has even developed four concentric "undulating membranes" on the periphery of the suction disc.

In addition to the suction cups, these microscopic animals may make use of hooks (one hook in Hoplidophrya, two in Steinella). Even in the protozoans, then, there are multiple attachment mechanisms like those which the parasitic flatworms have developed in great variety.

In the plant world there are no suckers in the true sense. However, roots can form "attachment discs" of similar appearance; their primary mechanism is probably gluing. Seedlings of the tropical epiphytic Loranthaceae, and the mistletoe of more northern latitudes, lock themselves to the host plant by means of a disc-shaped structure, the "adpressorium"; from its underside there is sent out a primary root (which acts as a sucking organ, or "haustorium") into the host tissue. In the Loranthaceae the lateral roots grow along the surface of the bark, here and there forming adpressoria which send out new haustoria (463). The nodular attachment "roots" of the wild grapevine and the rhizoids of some seaweeds, which have a similar appearance, should also be mentioned in this connection.

2.2.6.2 Double Suction Cups

The "bothria" or suction depressions of the cestodes (tapeworms) are found in pairs in certain orders; in some cases these exist as true pairs (Pseudophyllidea) and in others, as a consequence of paired fusion of four initially separate primordia (some Trypanorhyncha). Ruffled (464) or barrel-shaped (465) modifications occur.

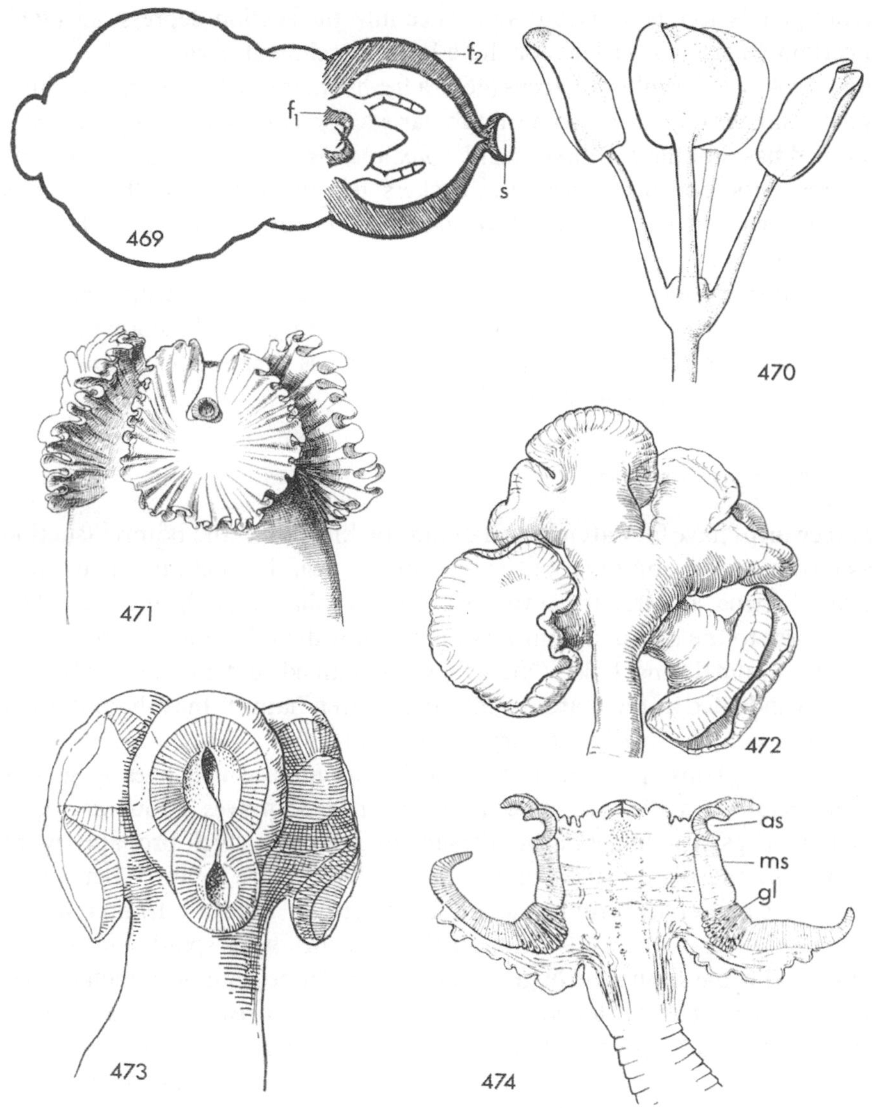

Plate 34 (Figs. 469–474)

469. Maxillary sucker of Achtheres percarum; f_1, f_2, first and second maxillae

470–474. Various forms of suction organs on the cestode scolex

Numerous glands secrete a viscous substance into the suction depressions (466), which certainly acts as a sealant for these large attachment organs.

The "carplouse" Argulus foliaceus (467), a fresh-water crustacean (Branchiura) barely a centimeter long, has two deep tubular suction cups (468). They represent the modified bases of the two first maxillae *mx* and always operate in synchrony. This ectoparasitic crustacean uses the suckers to attach itself to the surfaces of fish; only when it has found a favorable position does it bring its hooked clasping antennae into play.

The copepod crustacean Achtheres percarum (469) is provided with a peculiar suction apparatus. Here the ends of the enormously exaggerated second maxillae f_2 are joined at the tips to form a sucker *s* which is functionally unitary but arises from paired embryonic primordia.

2.2.6.3 Multiple Suction Cups

Many tapeworms have four attachment organs on the scolex: the bothria (suction depressions) are often multiple, but the bothridia (leaflike suction organs) and the acetabula (true suckers) are always so. The leaf-shaped (470) or fluted (471) forms conform to the irregular contours of the gut wall and follow the peristalsis of the gut without losing hold. Often they are stalked and mobile (472), and occasionally indented at the margin (473). It can even happen that the bothridia *ms* themselves contain small accessory suckers *as* (474).

The tarsier Tarsius spectrum is a lemur-like primate which lives in groves of smooth bamboo stalks on some asiatic islands. It moves about at night, making long leaps. The fingers have disc-like digital pads which probably work analogously to the digital discs of tree frogs. The mouse lemur Microcebus murinus, the smallest lemuroid, lives in Madagascar. Like all lemurs it has strong jumping legs and moves through the branches with speed and agility. In landing and while sitting the fingers and toes are fanned out with metacarpus and metatarsus bent at an angle of about 90° (475). On each end joint there is a prominent round cushion which acts as a sucker when the central surface is pulled back. If one tries to lift the animal from the substrate it becomes quite clear how strongly the fingers and toes are attached. Large, elongate pads are found on the fingers of the Philippine tarsier Tarsius syrichta (476). These organs are a useful aid in landing after the characteristic meter-long leaps through the bushes which these animals make (477). Digital pads are also present in the other lemuroids, though less well developed.

Six suckers are used by the rhagophilous larvae of the net-winged midges Blepharocera, Liponeura, Hopalothrix, and related genera of the Blepharoceridae. These flat larvae, which look like wood lice, live in very rapidly flowing mountain brooks. There they are found where the current is extremely strong—even in roaring waterfalls. They attach themselves to stones with the six suckers, which are oriented in a row along the midline of the body (487–490). Each

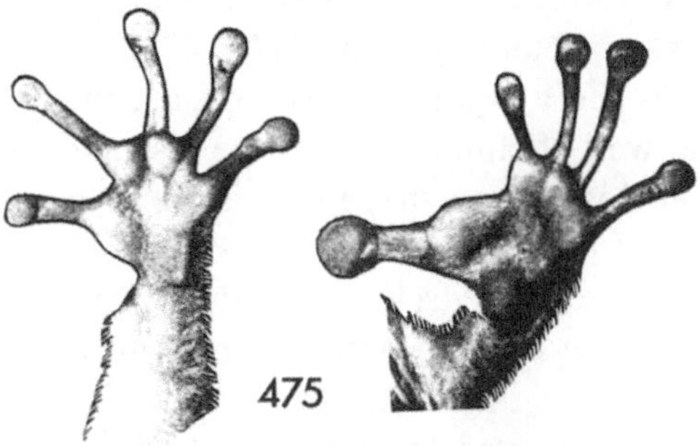

Plate 35 (Figs. 475–477)

475. Adhesive pads on the fingers and toes of Microcebus murinus, pressed onto a glass plate. Left, forefoot; right, hindfoot

476. Digital pad of Tarsius syrichta

477. Tarsius syrichta jumping

sucker is formed by a fold of skin and has a central chitinous cone which serves as a "pump piston" (478, 479). A system of slender chitinous ridges reinforces the margin. This sits on a conical stalk which opens outward. The external wall of chitin between the disc and the stalk is split into three lamellae which enclose a ring-shaped cavity filled completely with a secreted fluid. The pressure of this fluid evidently gives the suction apparatus additional stability. The larva attaches itself so firmly to the slippery surface of a stone that one can hardly get it off without tearing it to bits.

The delicate outer ring of chitin, with its microscopically fine longitudinal furrow, is supported by sword- and stiletto-like rays and has many crater-like depressions. The first step in attachment is to place it on the surface of a stone. The rim is well sealed by a dentate border of fine hairs and by a putty-like substance secreted from the tubular glands. Finally, dorsoventral muscles lift the central cone of chitin reinforced with firm radial struts, creating the required low pressure. The larvae can wander for meters over the stones on the bed of a stream, grazing on the surface coating of diatoms. They move like inchworms, alternately releasing and attaching the individual suckers, or by twisting from side to side, releasing and reattaching the three front suckers and the three back suckers in alternation. Differently constructed and less complex suckers (480, 481) are found on rheophilous larvae of some moth-flies (Psychodidae), which live in the waterfalls of jungle streams in Brazil, Java, and Sumatra. But their number and geometrical arrangement along the median ventral surface correspond exactly to the blepharocerid type.

These organs of the blepharocerid larvae are among the most remarkable attachment devices evolved. They are complicated and therefore have frequently been misinterpreted, with respect both to anatomy and to function. Recently, a careful study of the species Liponeura cinarescens has been made by P. Rietschel. Therefore the principles of the design cited above can be supplemented by a detailed discussion of fine structure and function of the suckers in this one selected species (482–485). Interpretations of previous authors which have been shown to be erroneous will not be mentioned.

Plate 36 (Figs. 478–485) ▶

478. Principal structures in the sucker of a blepharocerid larva
479. Median section through the sucker of (478)
480. Arrangement of suckers in a psychodid larva of the genus Hydropedeticus
481. Single sucker from (480)
482. Fine structure of the sucker of Liponeura cinarescens; axial section through the distal end
483. Section in the plane of the disc of (482), showing the cephalad sector with notch
484. Sucker with preformed replacement sucker, Liponeura cinarescens. Fourth larval stage just prior to molting. Labels with the prime symbol refer to the crumpled replacement sucker.
485. Schematic drawing of (484)

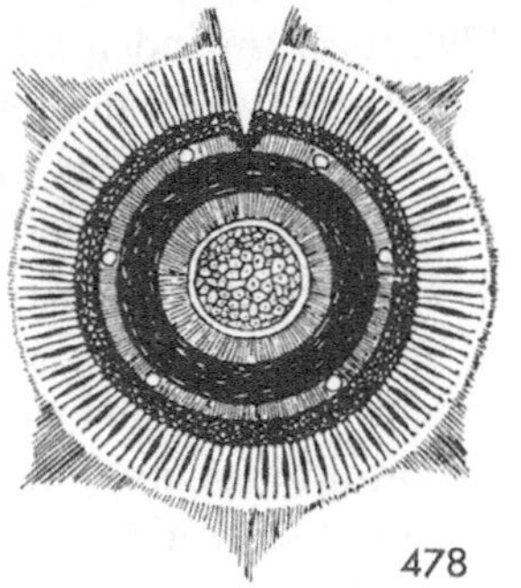

478

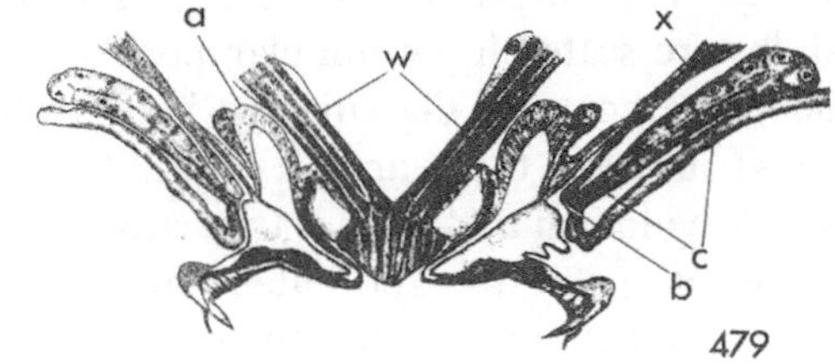

479

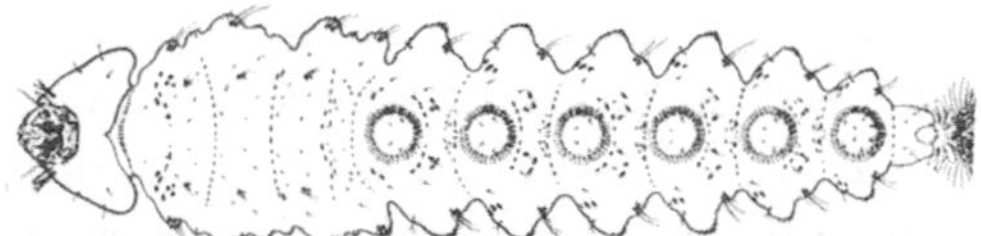

480

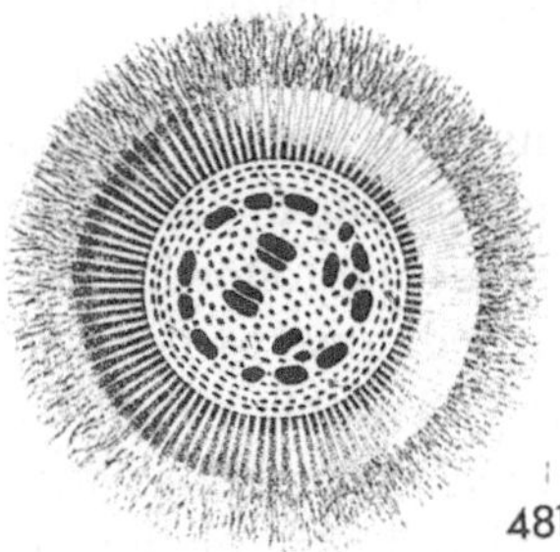

481

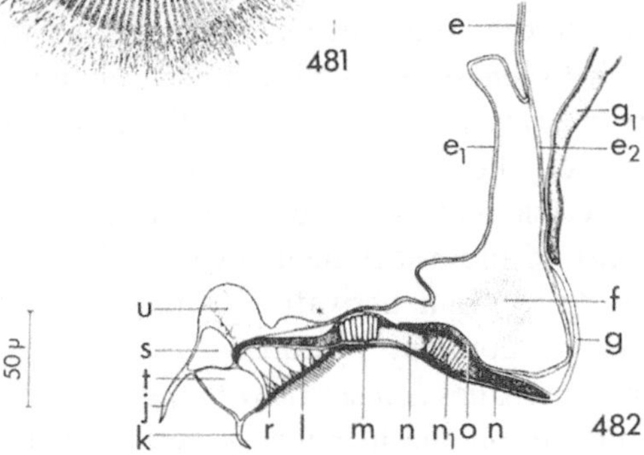

482

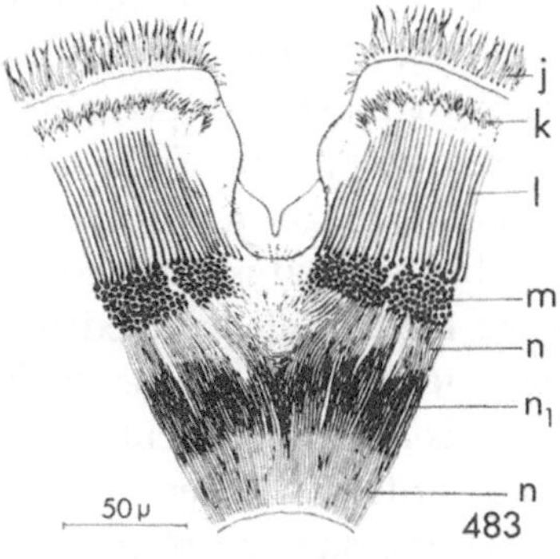

483

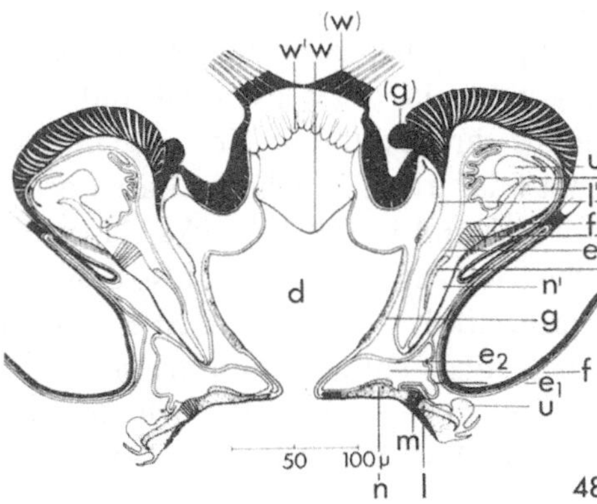

484

485

In Figs. 482, 484, and 485 the attachment organs are shown opened out. The stalks are seated in an annular pocket (cf. Fig. 479), the skin indentation *b*, which, together with the adjacent body wall, forms a ring of skin, the ring fold *a*. At its apex there are six large, tubular glands *c*. The stalk has the shape and the lumen *d* (484) of a truncated cone. At its upper end it is bounded by the piston apparatus. The outer wall of the stalk *e* splits distally into two lamellae, an outer one e_1 and an inner one e_2. These are the boundaries of an annular lumen which is completely filled with secretion *f*. The inner wall of the stalk *g* is strongly sclerotized along the inner two-thirds of its length. Since it has the shape of a truncated cone, this circumferential sclerotized zone forms a corresponding ring which may be called the "cuff" g_1· (482). At the proximal end the cuff is continued as a delicate, ring-shaped joint membrane which is connected on all sides to a central chitinous cone, the piston.

The piston is strongly sclerotized on the outer wall; tonofibrils which cross through the underlying epithelium transmit the pull of its three pairs of retractor or piston muscles *w* to the hard cuticular layer. The proximal part of the attachment apparatus as thus far described, then, includes the following basic functional units: first, a stable "cylinder", the cuff, with an additional hydraulically stiffened ring, the secretion-filled lumen. Second, there is a stable piston with the necessary withdrawal apparatus and force-transmitting connective elements. The third is the indispensable seal between piston and cylinder, in the form of a ring membrane. From a bioengineering point of view, all the components for production of a reduced pressure on the piston-and-cylinder principle are available. All that is lacking is a dependable seal to the substrate. This is formed by the highly differentiated distal part of the attachment organ which can be designated *in toto* as the "adhesive disc." As a derivative of the outer stalk wall, it is a purely cuticular formation. It represents a more or less flat washer-shaped structure. Its central, circular opening lies concentric with the cylinder-piston system, and the plane of the surface, on the average, is perpendicular to the axis of the cylinder. The margins are slightly fluted. A segment of a section in the plane of the disc (483) shows the cephalad interruption in the otherwise closed circular structure. Further details are shown in Fig. 484.

A large number of concentrically arranged elements serve in part for attachment and in part for stiffening. Taking them in order from the outside in, there is first a border of delicate chitinous threads, the upper marginal fringe *j*. Close within this is a second, somewhat finer fringe *k*. Next appears a system of radially oriented outer rays *l*, all of which end in tips bent like claws. They are alternately thin and thick, at least at the base, and therefore are also designated "stiletto bristles" and "sword bristles", respectively. Further inward is a narrow ring of "palisades" *m*. These are tall polygonal chitinous prisms stacked up like pillars of basalt. Their axes are parallel to the axis of the cylinder-piston system. The chitin micelles are also oriented in this direction. Next to the palisades, at the center, is the ring of inner rays *n*. Their number is greater than that of the outer rays. In contrast to the latter, these branch extensively and some of them end before they reach the palisade zone. They are shaped not like threads, but rather like lamellae, and are thickest in the middle of the zone n_1. The upper and lower edges of each lamella are especially strongly sclerotized. Here, as in the outer rays, the chitin micelles are oriented radially and thus lie perpendicular to the preferred direction in the palisades. Resistance to tangential stress (by zones *l* and *n*) and centripetal pressure (by zone *m*) are therefore combined. The thickest parts (n_1) are overlaid dorsally by a layer of circumferential chitin fibers of the ring-fiber layer *o*. At six points on the periphery of the *n*-zone the inner rays separate slightly; in each of the resulting clear fields one finds one or two sickle-shaped pointed hairs—withered cuticular structures.

The sole of the adhesive disc, in the region of the outer rays and palisades, is densely covered with microhairs. The tips of the hairs in this velvety covering are directed somewhat centripetally. Very much shorter microhairs cover the sole beneath the inner rays. All these structures evidently aid the spreading and mechanical stability of the disc, and facilitate sealing and fine-grain anchoring. In addition, there are other elements which may be described as ring reinforcers or stabilizers. Near the rim, the sole membrane of the disc separates more and more from the outer rays. Chitinous fan fibers *r* pass through the resulting nearly toroidal cavity and connect the ceiling and floor plates elastically. Further distal, a dorsal (*s*) and a ventral (*t*) ring canal surround the whole disc. Moreover, a marked enlargement *u* encircles the upper margin, interrupted only in the region of the cephalad notch.

There are six pairs of sucker muscles which run from the dorsal body wall to elements of the attachment organ. Three pairs go to the piston *w* and two, to the place where the ring pocket folds over, between the stalk and the ring fold *x*. Another muscle is inserted centrally, and two pairs of lateral muscles may be involved in the activity of the organ. All these muscles belong to the dorsoventral system. A muscular arrangement of this kind is diagrammed in its entirety in Fig. 486, for the Chilean blepharocerid genus Edwardsiana, which is built somewhat differently.

To continue the description of the sucker of Liponeura, there are three pairs of large tubular glands which open directly at the apex of the ring fold. It is likely that their secretion keeps the disc surface from being wetted by water and acts as a sealing compound for the marginal fringe. After the disc has been placed on the substrate, the piston (*w* in 484) is raised by the piston muscles. As long as these muscles remain contracted, and as long as the junction between disc margin and substrate is well sealed (the pliable elements of the outer edge follow closely the irregularities of the substrate), the resulting internal low pressure is effective. The larva remains firmly attached. The inner rays *n* complete the distal portion of the suction cyclinder. The lamellar construction of this zone, thickened and with extra support in the middle, keeps the disc from being deformed toward the center by the pressure difference. Around this zone, the palisades form a ring-shaped joint for the movable disc margin, which is reinforced by the outer rays. The V-notch (483, 485) probably adds to this mobility and, like a sort of escape valve, makes it possible for the water to flow out when the disc is initially pressed in place. Before the piston is raised to make the suction connection, the entire surface of the disc must of course be in contact with the substrate. This positioning of the disc must occur with the aid of the ventral longitudinal musculature of the larva, and probably proceeds from the inside outward. The fine brush hairs provide additional anchoring on the surface of the stones. To do this, they must penetrate the fragile gelatinous diatom coating, on which the larvae feed, and contact the surface of the stone directly. Without such a mechanical catch mechanism it is possible that the animal, together with its suction mechanism, would slide along over the gelatinous layer when subjected to strong currents. The same task is doubtless assigned to the brushes associated with the "claw appendages" which flank the attachment organs on the ventral side.

When the suction attachment is to be released, the inner cavity must be reduced in size. This can occur passively, by relaxation of the muscles *w*, and actively as well, by contraction of the muscles *x*. The existence of a directed-flow system by way of valves had been suspected, but no morphological basis for this could be found in Liponeura cinarescens.

The attachment organs of the Blepharoceridae can, as a whole, be retracted by means of the muscles of the annular pocket. As soon as the connection with the substrate is made they are pulled in. Hence the ventral wall of the larva is drawn flat against the substrate, so that the animal projects only slightly through the boundary layer into the full current. Shearing forces are thus much reduced. The suckers are probably re-extended by blood pressure; in any case, no antagonists to the ring muscles are present. More primitively constructed attachment organs of psychodid larvae (Horaiella, Sycorax, Marnina, Neotelmatoscopus) are unstalked and therefore not retractile.

Blepharocerid larvae wander slowly over the diatomaceous meadows on which they feed. At most, three of the six suckers may be released from the substrate at any one time. The others must hold firm, or the larva will be washed away. The same problem is encountered during molting. Most of these larvae molt three times before pupation. Since it is a cuticular structure, almost the whole sucker is molted. This situation has two consequences. First, the molting of the suckers must proceed in sequence, so that the new suckers are already attached before the old ones are released. Second, the new sucker must be preformed in a functional condition above the old one, like a spare part, and it must harden quickly. But since the space dorsal to the suckers is limited—the animal is very much flattened dorsoventrally—the newly-developing sucker can occupy only a small volume. These contradictory requirements are met by an ingenious gambit: this complicated structure is formed in a folded condition.

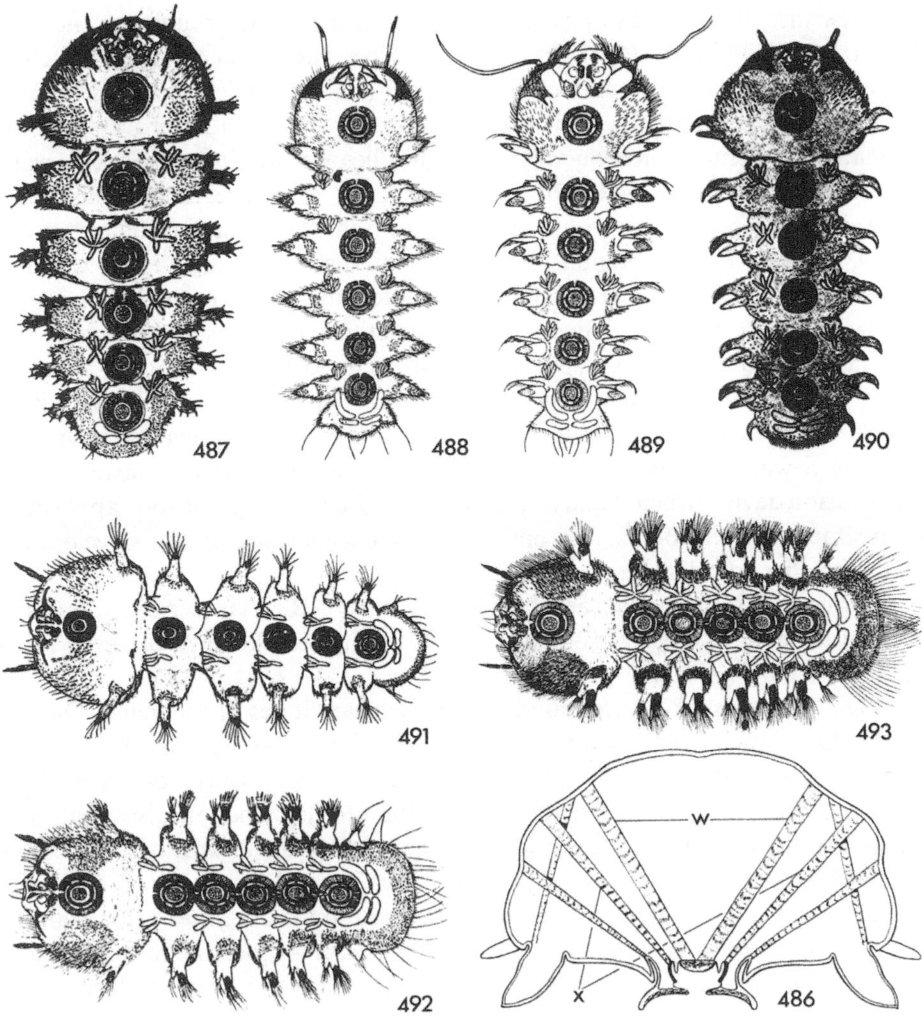

Plate 37 (Figs. 486–493)

486. Musculature of the sucker in the larva of the Chilean blepharocerid Edwardsiana spec.
487. Larva of Apistomyia tonnoiri (Blepharoceridae)
488. Larva of Blepharocera fasciata (Blepharoceridae)
489. Larva of Liponeura brevirostris (Blepharoceridae)
490. Larva of Bibiocephala spec. (Blepharoceridae)
491–493. Second, third, and fourth larval stages of Hopalothrix spec. (Blepharoceridae)

The new disc, with its rays already hardened, is arranged around the old stalk like a folded umbrella (485). In departure from the ordinary mode of chitin secretion, the peripheral part of the disc, with all its differentiations, is built by preformation of all the individual elements by cytoplasmic processes of the formative epithelium; subsequently, chitin micelles are laid down in the cytoplasm to transform it into cuticle. During molting, the structure thus prefabricated is rapidly pumped up to a functional condition. The same principle, as we know, is found in the imaginal molt of insects: here the wings are folded into the narrowest possible space and pumped up after emergence. In the blepharocerid suckers, a two-cell gland delivers a fluid secretion at a considerable rate. This is pumped into the narrow space (*f′*, 484) enclosed by the disc and diagonal inner lamella of the outer wall of the stalk, as well as the outer wall itself, which is folded like a bellows. This causes the bellows to expand and form the ring *f*, which is filled to rigidity by fluid. The intracuticular space is thus a distension device which deploys and stabilizes the disc. This curious application of hydraulic distension is just one of many peculiarities in the development of the blepharocerid sucker.

For comparison, the larvae of several species of Blepharoceridae are drawn in ventral view in Figs. 487–490. Figs. 491–493 are sketches of three larval stages of a single species.

Eight stalked suckers are found in monogenean flukes of the genus Diclidophora (494), which lives on marine isopods. Young whip-scorpions (Holopeltidia, Pedipalpi) have suction discs at the ends of the tarsi instead of claws. With these they hold themselves on the back of the mother. Suction discs with long stalks are also found on the males of the grain-mite Acarus siro (495), on the first, second and fourth pairs of walking legs, while the third legs each end in a trailing hair. Migratory nymphs of the sarcoptiform mites—for example, the flour mite Tyroglyphus farinae (496)—have a plate behind the coxae with several rows of suckers, as well as single suckers on the coxal surfaces. They use these to attach to the smooth cuticle of insects and isopods, and let themselves be carried about and hence distributed. Another mite, Antennophorus uhlmanni anchors itself with minute suckers, usually to the underside of the head of ants of the genus Lasius (497). When the carrier ant is fed by another ant, Antennophorus sticks its mouthparts into the droplet of food and takes some for itself. More remarkable, if the carrier ant is satisfied sooner than the parasite, Antennophorus couples the two ants together mechanically: it embraces the head of its host with the two pairs of hindlegs, while the second pair of legs, with the suckers, is attached firmly to the head of the donor.

Males of aquatic mites (Hydrarachnidae) have a ventral slitlike genital opening surrounded by a species-specific arrangement of 6 (498, 501) or many (499, 500) suckers. That these so-called genital fields actually function as suction devices, holding the partners together during copulation, has been demonstrated for the Arrhenuridae; presumably it is true for the other families as well.

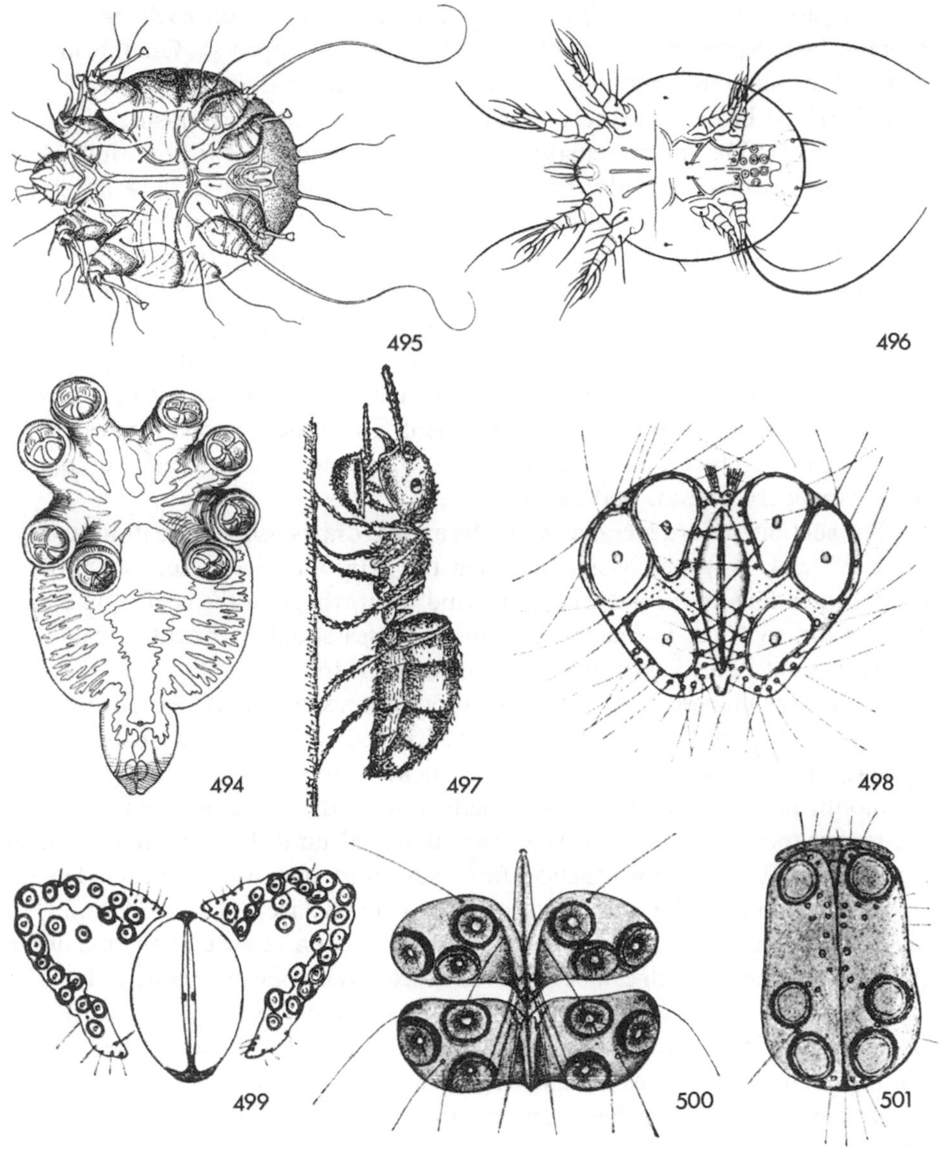

Plate 38 (Figs. 494–501)

494. Suckers of Diclidophora monenteron
495. Suction discs and trailing hairs on the legs of Acarus siro
496. Coxal suckers and trailing hairs in Tyroglyphus farinae
497. Antennophorus uhlmanni on a host ant
498. Suckers at the genital aperture of a male hydrarachnid, Hygrobates longipalpis
499. Suckers around the genital aperture of Piona rotunda
500. Suckers around the genital aperture of Atax crassipes
501. Suckers around the genital aperture of Limnesia undula

Not just a few suckers, but several dozen, are found on each arm of the dibranchiate cephalopods (squids, cuttlefish, and octopuses). On the arm of a large male octopus, an idividual sucker can be 4 cm in diameter. The octopuses (502, 503) have broad-based suckers, each with a flat, smooth, ring-shaped attachment surface (504). The musculature for effecting attachment is powerfully developed. The decapods (squids and cuttlefish), on the other hand, have developed hemispherical suckers formed by elaboration of a relatively thin stalk (505–511). The opening is always rimmed with a rigid, horny covering; typically this bears strong uniform teeth directed centripetally. In Figs. 512 and 513 the octopod and decapod suckers are compared, in slightly idealized median sections; the figure legend of 513 describes the structure in detail. If captured prey tries to free itself by violent movements, the hemispheres tend to expand, thereby actually strengthening the suction attachment. The hooked suckers are especially suited to the capture of fish and jelly-fish, while the smooth ones are adapted for grasping crustaceans and clams.

The suction forces generated when the animal draws back the central "activating rod" (514, 515) are considerable. On the skin of a diver, they can produce deep impressions of the sucker margin and hemorrhage in the center. The giant suckers (up to 15 cm in diameter) of the deep-sea squid Architeuthis can leave deep scars even on the skin of the sperm whale (516).

In the very much smaller species Octopus bimaculatus and Octopus vulgaris, these forces have been measured directly. Excised suckers of various sizes were attached to a spring balance and the sucker surface positioned on a piece of smooth polished wood. The musculature was stimulated electrically, so that the sucker took hold. Tension was then increased until the suction attachment was broken. The maximal suction force sustained by a single sucker depends upon the diameter; for diameters of 2.3 to 6 mm (areas 4.2 to 28.3 mm^2) it lay between *ca.* 30 and 150 kdyne (517). On the assumption that the maximum pressure difference which can be achieved at sea level is one atmosphere (1.033 kg-weight/cm^2 or 14.7 lb/in^2), the theoretical maximum force due to suction can be computed from the basal areas of these suckers. In all cases, these values are higher than the forces actually measured. The smallest suckers achieve up to 70% of the theoretical maximum force and the largest, only up to 45% (517). One can call this percentage the efficiency of the sucker. The values

Plate 39 (Figs. 502–509) ▶

502. Polypus herdmani (Octopoda)
503. Argonauta spec. (Octopoda)
504. Suckers of a cephalopod, attached to a glass plate
505. Proximal part of the tentacle of Mastigoteuthis cordiformis, with stalked suckers
506. Tentacular sucker (from 505), seen from the side
507. Tentacular sucker (from 505), surface view
508. Sucker from the arm of M. cordiformis, seen from the side
509. Sucker of (508), view toward the opening

502

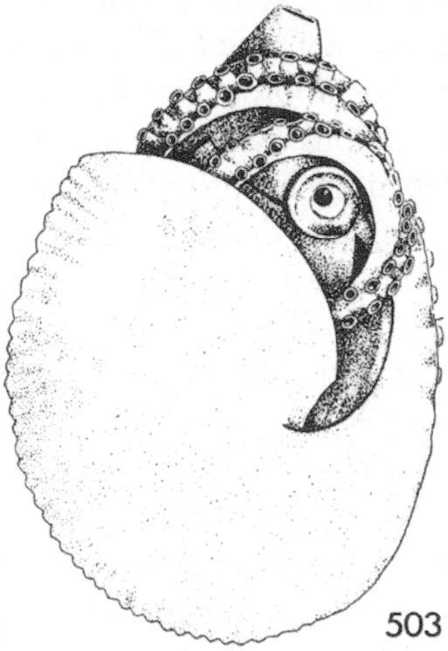
503

504

505

506

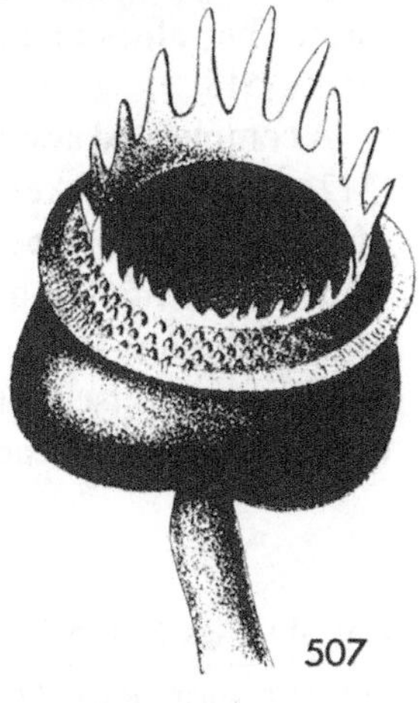
507

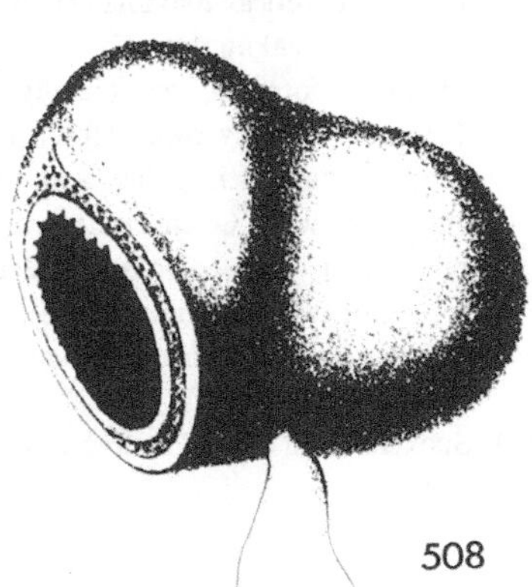
508

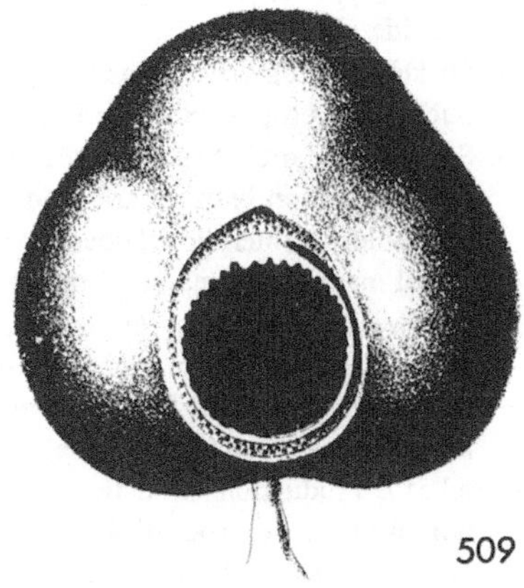
509

are consistent with those commonly found for suction devices in engineering. On the other hand, the sea anemone Cribrina can generate a pressure difference of one atmosphere and thus approaches the efficiency maximum of 100%. The force which an intact octopus can exert on an object may be measured by offering it a crab fastened to a spring balance and letting the octopus pull on the crab while holding on to the aquarium wall with some of its other arms. In this way one obtains an "average active tractive force" of the animal, which it can maintain for a considerable time. This never amounts to more than 50% of the "maximal tractive force" which the animal can develop for a short time—for example when one slowly pulls the crab away until the octopus lets go. The latter amounts to about 10^7 dynes (approx. 22 lb) for a medium-sized (0.4 kg-weight, or about 1 lb.) octopus. As shown in Fig. 518, the force increases with body weight more or less linearly by 2 megadyne per 0.1 kg weight, but only up to a body weight of about 0.5 kg. If one knows the number and diameter of the attached suckers, one can use the above-mentioned force measurements on single suckers to work out the expected total maximal force. The values predicted in this way agree quite well with the direct measurements (518).

A complicated system of muscles makes this high-suction performance possible (512, 513). The suckers of the octopus (512) consist of radial muscles, the fibers of which stand perpendicular to the cup wall, meridional fibers in the outer wall of both margin and cup, and ring fibers, which are particularly prominent at the opening. The connection between a fold of skin surrounding the margin and the sucker itself is provided by radial tracts of muscle. There is one connecting muscle bundle for each of four directions of movement of the cup (forward,

Plate 40 (Figs. 510–519) ▶

510. Tentacular sucker of Chiroteuthis imperator (Decapoda, Cephalopoda), seen from the side
511. Sucker of (510), view toward the opening
512. Sucker of Octopus; labels as in (513)
513. Sucker of the squid Sepia; c, circular fibers; con, connection between arm and sucker; g, ganglion; m, meridional fibers; ma, sclerotized margin; mr, marginal radial fibers; n, nerve; r, radial fibers; rm, radial muscles of the outer margin; rs, radial muscles of the sucker cup; smr, submarginal radial fibers; sp, sphincter; sf, marginal skin fold
514 and 515. Production of suction by retraction of the central rod; diagram
516. Sucker marks made by a giant squid on the skin of a sperm whale
517. Maximal suction (w) vs. surface area (a) for 8 suckers of Octopus bimaculatus. t.m.s., theoretical maximal suction; o.b.f., observed breaking force
518. Maximal force (f) exerted by different Octopus vs. body weight (bw). Filled circles: the animal is pulled away from the aquarium wall. Open circles: the animal releases a crab. Crosses: "detachment force" for two octopuses (0.15 and 0.395 kg weight) computed from the single-sucker measurements in (517)
519. Suckers of an adult Pneumoderma atlanticum

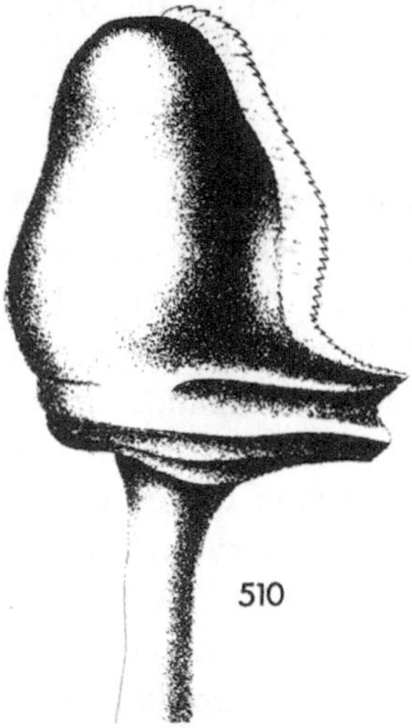

510

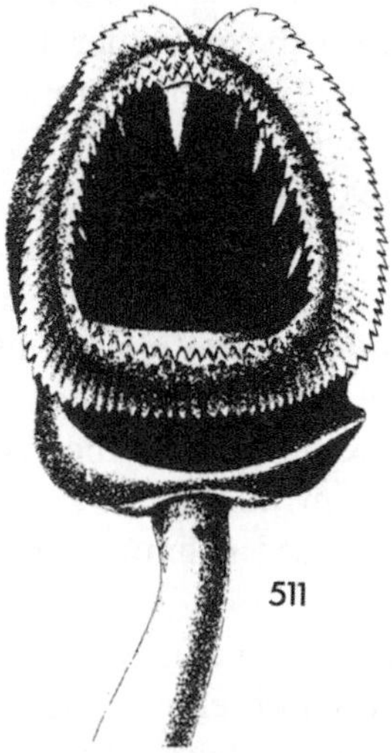

511

519

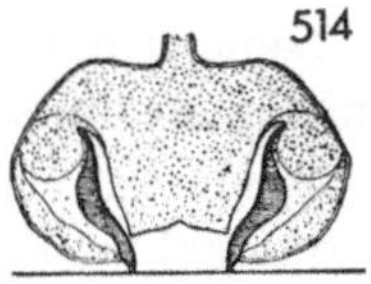

514

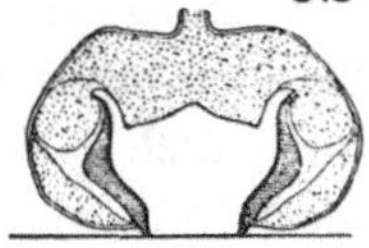

515

512

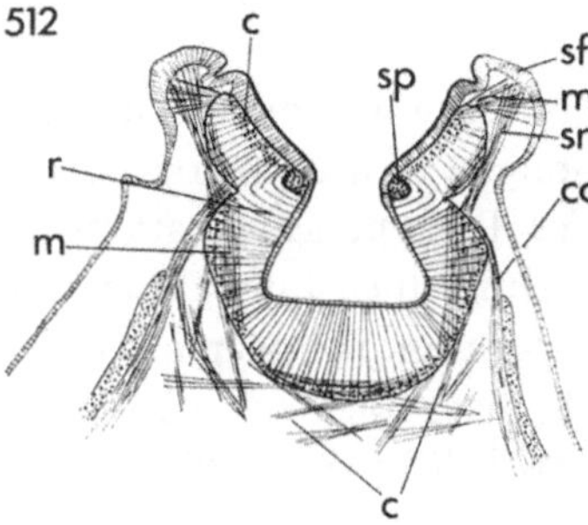

513

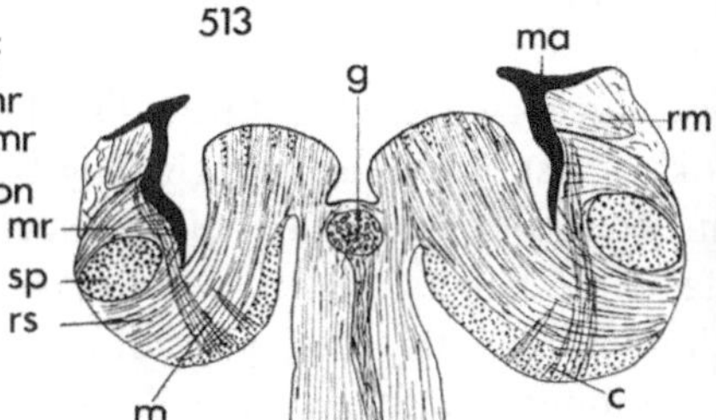

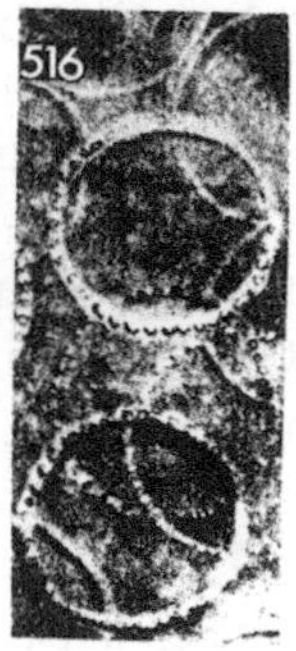

516

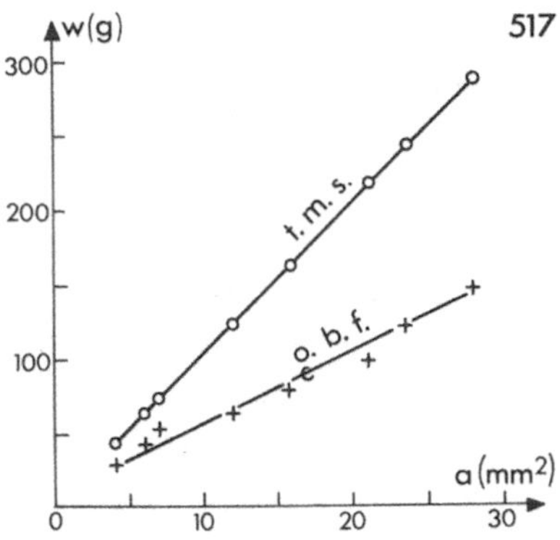

517

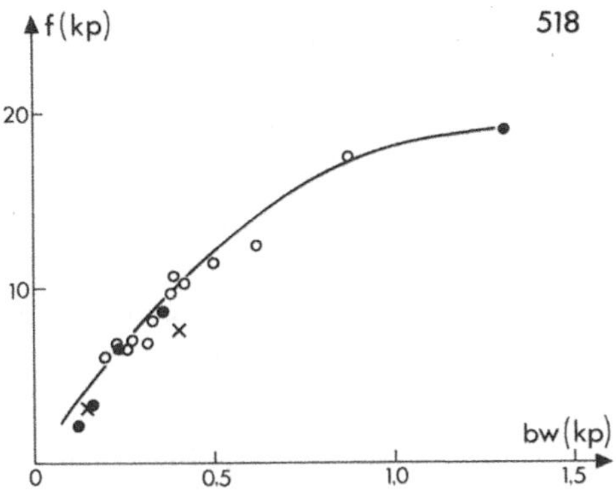

518

back, left, and right); each of these bundles inserts at one end on the outside of the sucker and at the other, on the muscular trunk of the arm. They are held together by a girdle of ring fibers, the contraction of which displaces the sucker.

When the sucker is set in place, the ring and meridional muscles are initially contracted. This action pushes the floor outward and reduces the size of the enclosed space. The margin of the sucker is flattened and pressed tightly to the surface. Now the seal is made by contraction of the ring muscles in the skin fold around the margin of the sucker. Then the cup expands. In this process all the muscles just named, except for the last-mentioned ring muscles, relax, but the muscles between arm and sucker, which dilate the outer wall, contract and thus increase the enclosed volume. Certain accessory muscles also act to enlarge the cavity.

The largest suckers are found in the genus Architeuthis. As mentioned above, suckers 15 cm in diameter have been observed. The smallest are probably those of the species Inioteuthis bursa, with diameters amounting to only 0.1 mm. Nevertheless, they are just as elaborately constructed as the large ones. Marginal teeth form an additional anchorage. Usually the number of such teeth varies inversely with their size. In the extreme case (Galiteuthis, Onychoteuthis, Abralia) there is but one giant hook-tooth per sucker. In such cases the suckers themselves are practically non-functional. The mode of attachment has changed from suction to hook-anchorage.

The armlike, sucker-covered outgrowths of the curious animal shown in Figure 519 look at first glance as though they belonged to a cephalopod. This, however, is an opisthobranch snail, of the species Pneumoderma atlanticum (Gymnosomata). The suckers are on long stalks. Neither their function (capture of prey? attachment?) nor the mechanics of their operation is understood.

Among insect imagoes, the suction discs of Dytiscus are the best-known and largest, as well as the most finely differentiated. The outer discs of the smaller suckers sit at an angle on their stalks. The arrangement seems to be adapted to the slant of the thoracic shield in the female. Twelve other European genera of the water-beetle family Dytiscidae also have such organs. The suction devices of the males of Hyphydrus are round (520, 521); on the two anterior pairs of legs there are no less than 3700 of these minute suckers. The diameter of the discs varies between 5 and 27 μm and the length of the stalks amounts to 40–67 μm. Similar apparatus is found on the tarsi of Hydroporus. By contrast, Noterus bears only one relatively large cup on each of the first three segments of the first pair of legs (522), and on the first and second segments of the second pair. Agabus represents an intermediate form, with 15 medium-sized suckers (523) per leg, making a total of 60. Eunectes, with 250 small and 2 large suckers per leg, is rather reminiscent of Dytiscus, as is the medium-sized swimming-beetle genus Acilius, with many small suckers (20 μm diameter), two medium-sized suckers (300 μm), and one large sucker (500 μm) on the much

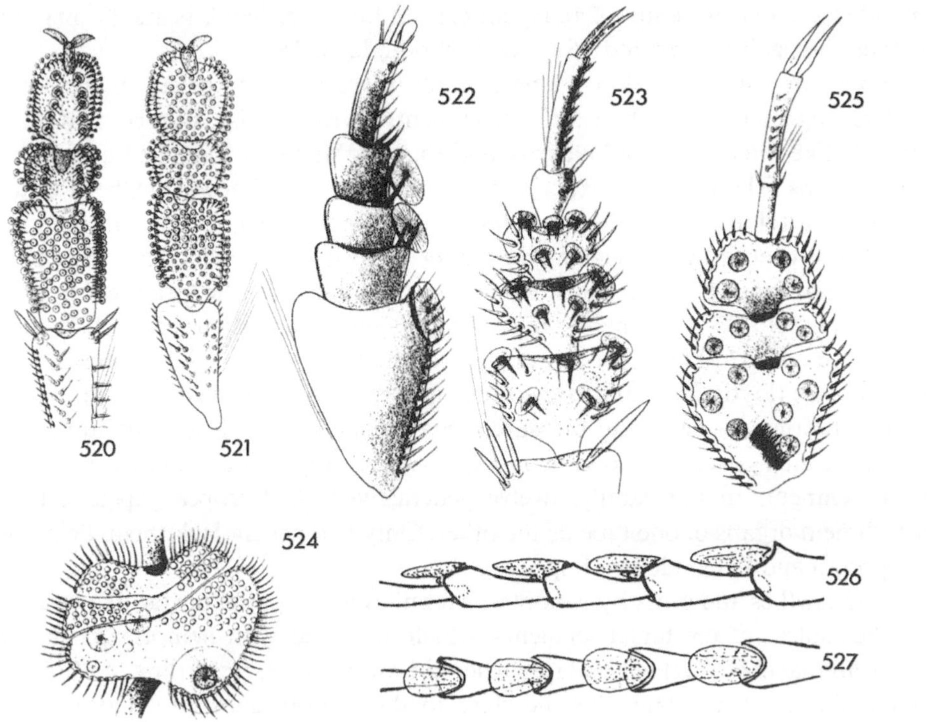

Plate 41 (Figs. 520–527)

520 and 521. Suction apparatus on the first tarsus of Hyphydrus ovatus

522. Suction apparatus of Noterus semipunctatus

523. Suction apparatus of Agabus agilis

524. Suckers of Acilius sulcatus

525. Suckers of Hydaticus hübneri

526 and 527. Tarsal vesicles of Cimbex connata, surface and side views

broadened anterior tarsus (524). Hydaticus, on the other hand, bears 20 smallish suckers on each foreleg and 15 on each middle leg (525).

One can see that within a single family there is great variety in the form and arrangement of such sexual attachment devices. It is assumed that they operate like the suckers of Dytiscus, and they are therefore classified as suction mechanisms. The indicated experiments have not, to my knowledge, been done.

A close relative of Dytiscus is the swimming beetle Cybister. Its anterior tarsus, however, bears no suckers but rather longish leaflets which are extremely pliant and flat. These certainly would be of no use in the production of a partial vacuum. Very probably, they cling by simple adhesion (see p. 150), just as do the comparable structures of the water beetles Ilybius, Laccophilus, and Colymbetes (Dytiscidae) and some other genera. If these leaflets are really true adhesion organs—again, no measurements are available—it is interesting that closely related genera of the same family have developed such different methods of attachment. In this family, twelve genera, with 98 European species, bear attachment organs of one type or the other. Only two genera lack them, Pelobius (1 species) and Cnemidotus (1 species).

The sawflies and other hymenopteran families have special blister-like vesicles on the "soles" of the tarsal segments, which are filled with hemolymph in the living animal (526, 527) but collapse when the animal dies. According to observations of Cimbex connata, they lie close to the smooth substrate. The vesicles secrete no fluid, but immediately after they have been placed on the substrate they show a ventral (usually irregular) indentation. Evidently this indentation is brought about by a decrease in the blood pressure. The tarsal vesicles thus appear to be capable of a suction-cup-like action. They can easily hold the animal in place, even without the assistance of the empodium. For sawflies, which run about a great deal on smooth leaves, these vesicles are certainly an advantageous contrivance.

The arms of the sea stars and sea urchins end in a number of suckers. The principle of their operation involves two antagonistic muscular systems: disc elevators, which insert in the middle of the strong connective-tissue disc and are connected primarily to the ambulacral plate of the calcareous skeleton, and disc depressors which run radially on the underside of the suction disc from its thick periphery to the thin center (528–531). When the elevators contract they raise the thin center of the disc away from the substrate, enlarging the central cavity and generating the reduced pressure necessary for firm attachment. But when the elevators relax and the depressors contract, the disc is again flattened; the pressure difference returns to zero, so that the whole podium may be lifted from the substrate by the contraction of longitudinal muscles.

Even the smallest echinoderms may have, in addition to other types of podia, those with well-developed suckers. The sea urchin Echinocyamus pusillus averages only 7 mm in length. The individual podium, when relaxed, is 140 μm long and measures only 35 μm in diameter. The pressure generated by the

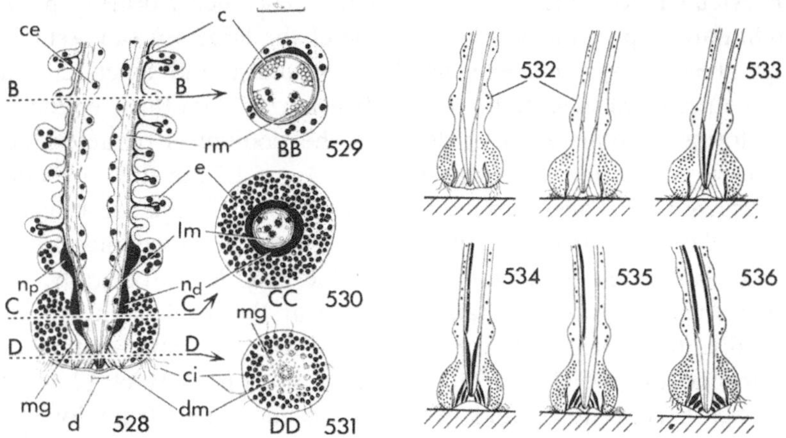

Plate 42 (Figs. 528–536)

528. Functional design of a suctorial podium of Echinocyamus pusillus

529–531. Sections through the podium of (528) at different levels

532–536. Model of the operation of the suctorial podium of Echinocyamus pusillus

ampullae can extend it to a maximum of 2000 µm. The sucker-bearing podia of this sea urchin are in principle similar to those of the large sea star Asterias rubens, though they diverge in some details of structure and function. The histological structure (528) is as follows: just beneath a 3-µm-thick epithelial layer *e*, which forms circumferential folds when the podium is retracted, lies a 1-µm-thick nerve plexus n_p, n_d. Within this is a 1- to 1.5-µm-thick cylinder of connective tissue *c* which consists exclusively of ring fibers. To this are attached the elevator muscles *lm* of the disc diaphragm *d* and the radial muscles of the disc *dm*, which spread out diagonally onto the disc. More centrally, the strong retractor muscles *rm* of the stalk terminate at the level of the sucker expansion. Finally, the podium is lined internally with coelomic epithelium *ce*. In a ring around the radial disc muscles, there are openings of 12 to 15 mucus glands *mg*. A concentric band of cilia *ci*, each about 10 µm long, is found near the periphery of the disc.

The attachment-movement-detachment process has been observed directly under the microscope. The podium is extended hydraulically by contraction of the musculature in the wall of the ampulla; the sensory cilia touch the substrate (532) and consequently the disc elevators contract. The diaphragm, the central part of the suction disc, is thereby pulled upward (533), so that the pressure in the contact zone is reduced. The mucus glands squeeze a sticky, sealing secretion into the ring-shaped contact zone. Now the disc muscles contract and further enlarge the cavity (534).

The stalk retractor muscles are arranged in 4 bands which contract independently and thus can bend the podium in an arc in any direction (535). When the substrate is firm this action produces a movement of the sea urchin and when it is loose, movement of a grasped particle relative to the animal.

The first step in detachment is relaxation of the disc elevators (535). The radial disc muscles, on the other hand, contract maximally and thus tend to make the diaphragm convex outward. In this way, proceeding from periphery to center, the disc rolls away from the substrate, which is sticky with mucus (536). Then, by contraction of all four bundles of retractor muscles, the podium is pulled up. Contracting muscles are indicated by the dark shading in Figs. 532–536.

The suction forces are large; for example, just ten suckers can support a 250-g sea star in air. In opening clams, their prey, large sea stars can develop tensile forces of up to 4.5 megadyne, which are transferred to the two clamshells by way of the several hundred suckers on the five arms. The medium-sized Asterias forbesi can exert maintained tensile forces of 3.1 megadyne, which are enough to pull the shells of small Mytilus edulis a millimeter or so apart. This small crack suffices for the sea star to introduce its stomach. Pull is evidently applied periodically, and rather quickly exhausts the closer muscle of the mussel. But the sea star is by no means exerting itself maximally. Since the tensile force of the retractor muscles of one podium amounts to about 30 kdyne (the

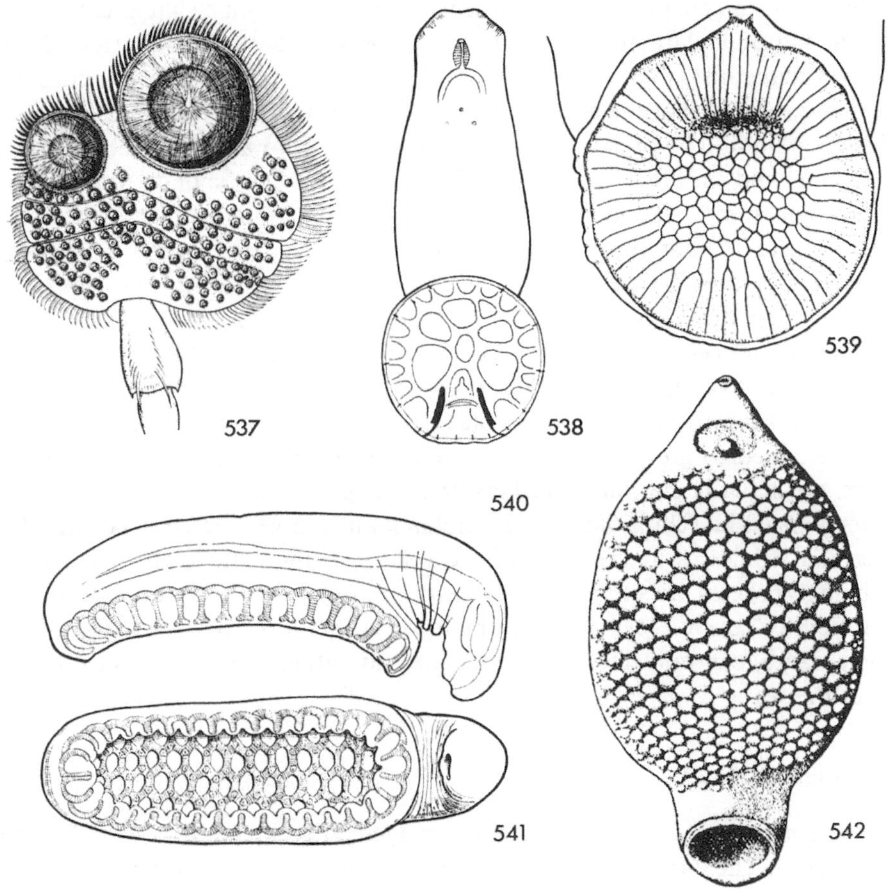

Plate 43 (Figs. 537–542)

537. Miniature suckers of Dytiscus marginalis; ventral view of part of the right anterior tarsus
538. Subdivided sucker of Merizocotyle diaphanum
539. Posterior sucker of Haemadipsa zeylanica
540 and 541. Ventral adhesive disc of Lophotaspis vallei
542. Adhesive papillae of Homalogaster

yield point of the suction attachment must be at least somewhat greater than this) one can calculate that Asterias need contract only one quarter of the podia simultaneously in order to develop a brief pull of 5 megadyne. Experiments have demonstrated that the suction attachment can in fact hold against tensions of this magnitude. The podia of Asterias vulgaris have an average sucker surface of 1.66 mm^2 and on the average can hold 29.4 g. This corresponds to a suction of 17.7 kdyne/mm^2. If one eliminates the suction action by boring holes in the test object so that the pressure differences between the sucker cavity and the exterior cannot be maintained, the podium still sticks to the substrate. Now, however, it can carry a maximal load of only 13 g, or 7.8 kdyne/mm^2. This residual adhesion is due primarily to gluing; the production of the sticky secretion has not yet been eliminated experimentally. Thus the suction mechanism itself accounts for about 56% of the total force of attachment and the remainder may be ascribed to gluing. The podia, although ordinarily described as typical suckers, are actually combined suckers and gluing devices.

Members of the Fibulariidae, a sea-urchin family, have differently constructed sucker-like structures at the ends of the podia; in these the only part of the attachment process accompanied by active muscular pull is the withdrawal of the slightly concave center. The brachiolaria larvae of some sea stars have a single, large, centrally located suction disc for attachment to plants on the floor of the sea.

2.2.6.4 Suction Plates

This term may be applied to suction arrangements extending over relatively large surfaces. They arise either by the juxtaposition of many individual suckers in a plane or by the enlargement and subdivision of a single attachment organ during the course of embryonic development.

An example of the first alternative is the anterior tarsus of the male Dytiscus marginalis. In these aquatic beetles there is, in addition to the two large suckers already discussed (p. 86), a field of on the average 138 close-set miniature suckers (537). The field as a whole can conform to the contours of the sharply arched thoracic shield of the female; each miniature sucker is so small, only about 0.1 mm in diameter, that it encounters negligible curvature. The total suction force of all 1880 suckers on the two anterior and middle tarsi of the male, under water, amounts to about 3 kdyne (comparable to the live weight of a female) when the attachment is to the back of the female, and 15 kdyne (about half an ounce) when the attachment is to smooth glass.

The monogenean fluke Merizocotyle diaphanum (Trematoda) has a posterior suction plate which has evolved by the subdivision of a single sucker into smaller chambers (538). This worm lives on rays, the skin of which allows only an especially closely fitting suction organ to be anchored reliably. The powerful posterior sucker of the leech Haemadipsa zeylanica is also subdivided

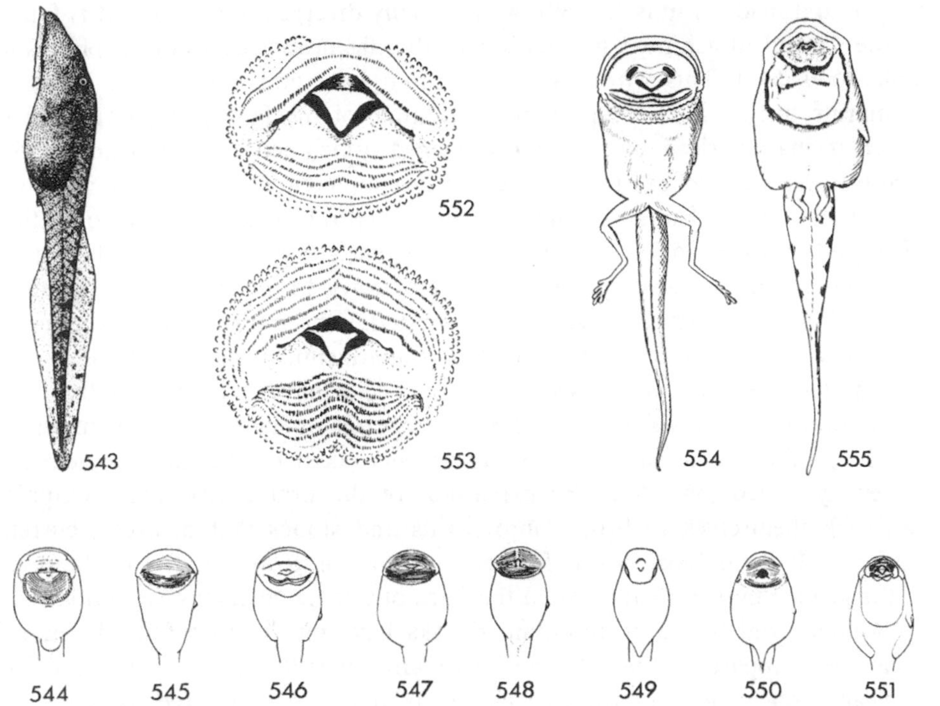

Plate 44 (Figs. 543–555)

543. Tadpole of Heleophryne natalensis, with "suction mouth"
544–551. Larval amphibia which inhabit flowing water on various continents, showing adhesive devices
552. Circumoral region of Hyla dominicensis
553. Circumoral region of Hyla heilprini
554. Tadpole of Bufo penangensis
555. Tadpole of Rana jerboa

by ribs and fields of muscle (539). An especially divergent development is found in the ventral attachment mechanism of the digenean trematode Lophotaspis vallei (540, 541). This worm is about 12 mm long and lives in the stomach of marine turtles. It is a typical representative of the family Aspidogastridae, which is characterized by a variably formed attachment disc in place of the terminal ventral sucker. The embryonic development of this organ indicates that it has arisen from a simple posterior ventral sucker by the formation of secondary longitudinal and transverse ridges. The result is an arched plate of many suction pits, arranged in three or four rows. In the fluke genus Homalogaster (Paramphistomidae, Digenea, Trematoda) the entire ventral surface is covered with papillae which participate in the attachment process (542).

The tadpoles of a whole series of frog species have "suction mouths": the circumoral region (involving both lips to a greater or lesser extent) is modified for the production of suction (543). The forms taken by this development are extremely varied (544–551). By expansion of the oral cavity, these tadpoles can suck themselves so firmly onto plants and stones that a strong current cannot pull them loose. Even further insurance is available to the tadpoles of the genus Staurois (Ranidae), in the form of a large accessory ventral sucker.

Animals which live in mountain brooks have much more fully developed attachment apparatus than do the inhabitants of still waters. The distinction is clear if one compares the two closely related species of tree frogs on the island Hispaniola: Hyla dominicensis (552) lives in quiet ponds and Hyla heilprini, with its remarkably developed circumoral region (553), in the upper reaches of the mountain streams. The rhagophilous tadpole of the Indo-Chinese toad Bufo penangensis (554) has an enormously convoluted circumoral region, with rows of teeth. Some frog tadpoles, for example those of Rana jerboa (555) and Rana hainensis from Java, have additional large attachment discs behind the oral region which are bounded by marked lateral swellings. Comparing these amphibians with fishes adapted to life in currents (e.g., Discognathus), one has an excellent example of convergent evolution.

Several species of fish have quite characteristic suction organs. They are located either ventrally or dorsally. The two closely adjacent pelvic fins which form a closed fan in the gobiids (556) have in the snailfishes (Lipariidae, 557) been transformed into a well-developed suction disc. The Gobiesocidae, on the other hand, have evolved a ventral suction organ by the transformation of bony elements—the claviculae, postclaviculae, and pelvic girdle; in Gastromyzon (558) the entire ventral surface, together with the pectoral and pelvic fins, serves as a suction disc. Behind the mouth of the carp-like Discognathus lamta (559) from Abyssinia, there is a broad suction-cup-like disc which ends anteriorly in a long swelling covered with papillae. Pseudecheneis sulcatus (560) has a transversely ridged attachment disc in the throat region. Occasionally even parts of the lower lip are modified into a suction cup, as in the cyprinid genus Garra (561) and the Indian species Orcinus sinnatus.

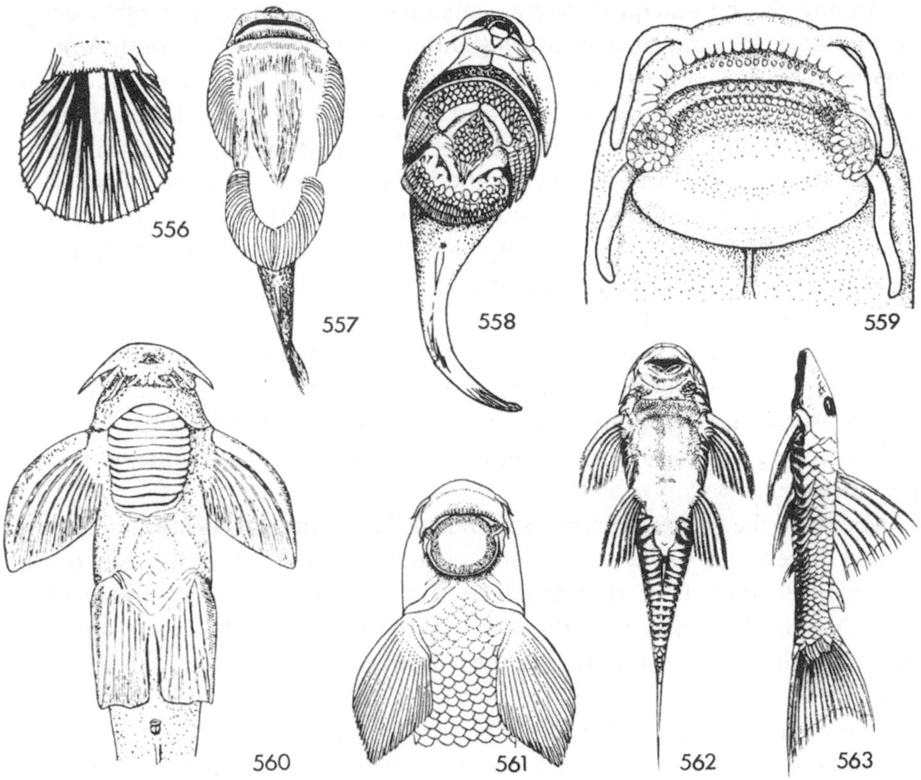

Plate 45 (Figs. 556–563)

556. Modified pelvic fin of a goby
557. Ventral suction organ of a snailfish
558. Suction disc of a gobiesocid fish, Gastromyzon spec.
559. Suction disc of Discognathus lamta
560. Suction disc of Pseudecheneis sulcatus
561. Sucker formed from the lower lip of Garra
562. Lip sucker and abdominal papillae of the armored catfish Plecostomus angipennatus, ventral view
563. Sucker and papillae of Plecostomus, side view

Among the best-adapted piscine inhabitants of mountain streams are the carp-like fish (Cyprinidae) and catfish (Siluroidea), especially the South American armored catfish (Loricariinae). In the Andean streams one finds genera such as Pygidium, Plecostomus (562, 563), and Arges, dorsoventrally flattened animals with broad heads and flat ventral surfaces. These fish use the lip region of the mouth, which is located beneath the body, to suck themselves firmly onto stones with the head directed upstream; for this purpose they are provided with dense arrays of attachment papillae (562). Even in rapids and waterfalls they manage to remain securely anchored. Such fish never swim for more than a short distance at a time.

The awkwardly built marine fish Cyclopterus attaches itself with its ventral sucker to stones in food-rich areas and simply waits, so to speak, for food to be washed into its mouth. As an example of the suction force exerted, a pull of 39 megadyne (86 lb) was required to detach an 8-inch individual from the substrate.

Males of the genus Kneria (Kneriidae) from Angola constitute a special case. In these fish, a sucker on the gill cover (operculum) (101, 564) is surrounded by a subcutaneous fiber ring *fr* which sits on a bony annular ridge of the operculum *o*. Together with the "postopercular organ" described on p. 25, this "opercular organ" serves to lock the male and female together during copulation.

The suction devices of fish described so far have all been located ventrally. Better known are the dorsal suction plates found on the anterior part of the head of the Echeneidae (remoras or sharksuckers; 565). These organs are extraordinarily powerful and superbly adapted to their function. The name of the order, "Discocephalidae", reflects this curious adaptation. It appears as an oval, flat depression, the floor of which is divided into sections by a double row of diagonally oriented, finely-toothed lamellae. Different species have from 10 to 28 pairs of lamellae. Ordinarily these lie flat, with the edges pointing forward. When the fish proceeds to lock onto an object, it shoves itself forward while touching the object. This pushes the lamellae upright, increases the total cavity volume, and lowers the interior pressure. A schematic illustration of this attachment process is given in Figs. 566 and 567. The enlargement of the cavity is of course due to the fact that the area of a rectangle is larger than that of any parallelogram with identical sides (568). To release itself, the remora simply moves backward, so that the lamellae fold down again. From an engineering point of view, an imitation of this mechanism ought to be easy to construct; but the author does not know whether the principle has ever been applied.

There are eight species of discocephalans, some of which can be up to 1 m long. Native Polynesians use them to catch turtles; they fasten the fish to a line and let it swim. When it finds a turtle it attaches itself and both animals can be pulled in.

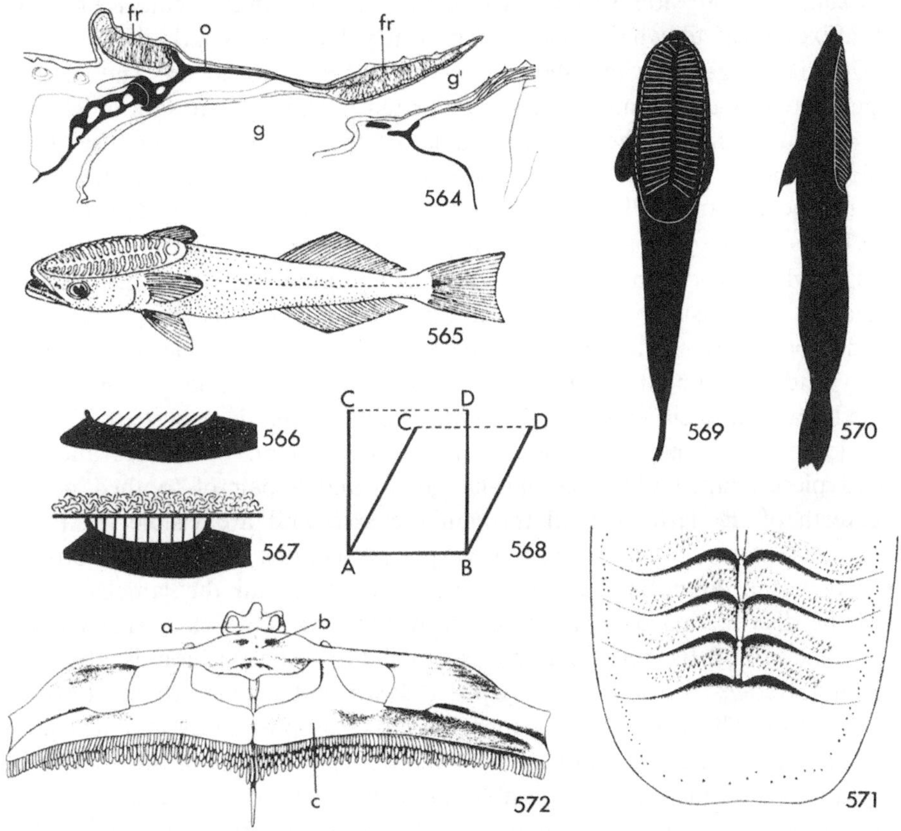

Plate 46 (Figs. 564–572)

564. Longitudinal section through the opercular sucker of Kneria spec.; g, gill chamber; g′, external branchial aperture
565. Echeneis remora, Discocephalidae
566–568. Attachment process in the Echeneidae, schematic

569 and 570. Remilegia australis, dorsal and side views; the slits in the suction disc are shown in white
571. Posterior half of the suction disc of Echeneis osteochir
572. Bony elements in a lamella of the suction disc of Echeneis naucrates

The suction depression is as wide as the fish itself and in the genus Remilegia, which affixes itself to whales, can reach nearly half the length of the body (569, 570). The organ is a modification of the anterior dorsal fin. The suction force generated by the Remilegia disc is indicated by scars in the skin of whales, which obviously mirror the outline and lamellar structure of the disc. It is possible of course that these scars arise secondarily—through infection and subsequent healing of small wounds produced by the lamellar teeth. The discocephalids Echeneis osteochir and Echeneis naucrates have an especially large suction plate (571) which has been the subject of thorough anatomical and histological investigation. It consists of 17 pairs of lamellae. (A fossil discocephalid with a considerably shorter plate and only six pairs of lamellae has been described; evidently the attachment mechanisms of the Echeneis species represent an advanced state in their evolution). Within each lamella is a system of four ossicles (572)—an unpaired spine-like ventral piece *a* on which lies another unpaired piece shaped like a clothes-hanger *b*, and a pair of toothed pieces *c*. The teeth of the latter extend through the skin and are visible from the outside. These are not true teeth, but simply projections of the bone. Strong connective tissue connects the basal bony elements of all the lamellae. The edge of the disc is ringed by a robust, movable rim in which a ring-shaped cushion of cartilage-like elastic tissue is embedded. The cushion is attached to the lateral edges of the hanger-shaped ossicles and extends to the outside of the disc rim. The whole arrangement forms a supporting system which is elastic but at the same time quite stable. Countless goblet cells in the epidermis secrete the mucus which is indispensable for such a suction apparatus.

The musculature of the disc is complicated. Three primary groups of muscles may be distinguished: (1) discal muscles between the individual bony structures of the disc, (2) capitodiscal muscles between the skeleton of the head and the disc ossicles, and (3) cervicodiscal muscles between the lateral musculature of the trunk and the disc ossicles. Each of these muscle assemblages can be subdivided into anatomically and functionally differentiated subgroups. To go further into the kinematics of this process would be to stray too far from our present purpose. In summary, though, the muscular system enables relative displacements of the bony elements which ultimately cause raising or folding of the lamellae.

Fundamentally, the establishment of suction requires that all the paired, toothed elements *c*, which are slanted forward at rest, be raised to a more vertical position. In fact, they are pulled by muscular action somewhat past the vertical, through the "top-dead-center" position, until, angled slightly back, the toothed processes bump against the lateral processes of the hanger-shaped ossicles *b*, which act as stops. The muscles which have brought them to this position can now relax, for with this orientation the suction tending to collapse the cavity makes the system intrinsically stable. In detachment, the releaser muscles need only rotate the toothed ossicles back through top-dead-center. Beyond this point, suction acts to fold the lamellae until they touch like shingles.

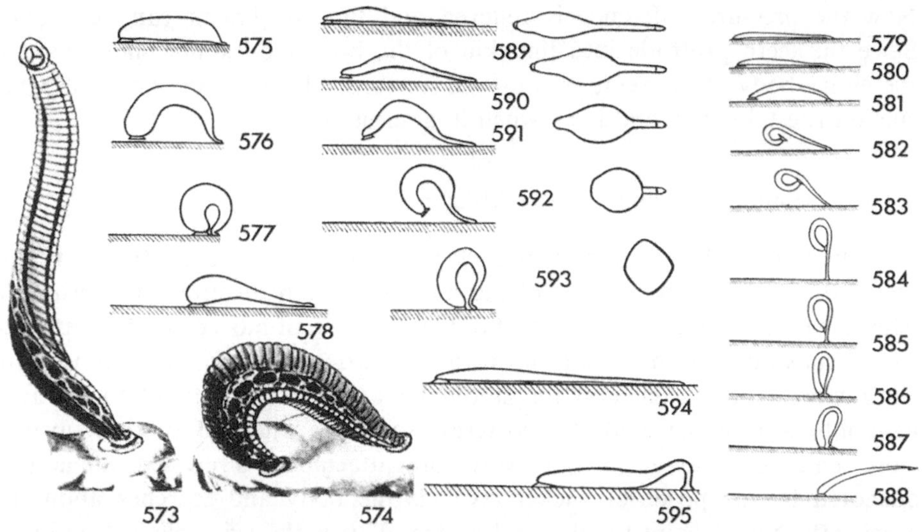

Plate 47 (Figs. 573–595)

573 and 574. The medicinal leech Hirudo medicinalis, with a sucker at each end

575–578. One possible sequence of creeping movements of Protoclepsis tessellata

579–588. Loop formation during locomotion, Piscicola geometra

589–593. Lateral and dorsal views of Helobdella stagnalis during locomotion

594 and 595. Ordinary locomotor movements of Herpobdella atomaria

Now the pressure difference is relieved and the attachment can be broken. Since the teeth protrude into the skin of the host and cannot slide along it, the entire body of the remora is moved forward relative to the host fish when suction is established, and back when it is released.

2.2.6.5 Locomotion Using Suction Attachment

For this application, at least two suckers, at different sites on the body, are required (573). The locomotion of the medicinal leech illustrates the principle. After the oral sucker, which is aimed in the direction of movement, has attached to the substrate, the posterior sucker is released and drawn up behind it by contraction of the segmental longitudinal muscles. At this point the body is bent into a U-shape, and the posterior sucker is attached to the substrate close behind the oral one (574). Now the oral attachment is released; the animal, anchored by the posterior sucker, extends the body and searches about for a new attachment point for the oral sucker. This is the principle of "leechlike" locomotion. Functionally, it is identical to the "inchwormlike" locomotion of some lepidopteran and sawfly caterpillars, as well as to the "wormlike" locomotion of many worms. In all these cases alternate anchoring of opposite poles of the body is involved. In the first example this is done by means of suckers, in the second by grasping hooks on the legs, and in the third by swelling (expansion-anchoring) of part of the body. Not all leeches behave like the medicinal leech in drawing up the posterior sucker. For example, Protoclepsis tessellata (Glossiphoniidae; 575–578) slides the posterior sucker forward along the ground. The fish leech Piscicola (Ichthyobdellidae), on the other hand (579–588), bends the posterior end of the body into a small loop, raises the body almost vertically and then slides the posterior sucker down along the ventral surface of the body until it attaches next to the front foot. An intermediate form of locomotion has been developed by Helobdella stagnalis, illustrated in both lateral and dorsal view in Figs. 589–593. The posterior sucker is raised somewhat from the ground and brought close to the body wall, but shortening is limited to the posterior portion of the body. Still another kind of locomotor coordination can be observed in Herpobdella atomaria (594–595). Here the whole posterior half of the body is pulled along, and it is the front half that is arched up and forward.

Hemiclepsis marginata decides upon a new direction of locomotion by making search movements with the anterior end of the body, which is extended, raised high and swung about like a whip while the posterior end remains attached. This species can even use the surface of water as a substrate. On a very loose substrate, the Ichthyobdellidae cannot progress at all, the Glossiphoniidae can barely do so, but the Herpobdellidae manage quite well. The Haemadipsidae, terrestrial leeches of the tropical zone, employ the two-sucker principle to climb bushes and trees. From this vantage point they drop onto a suitable host when it passes below and vibrates the substrate. These 3- to 8-cm-long parasites

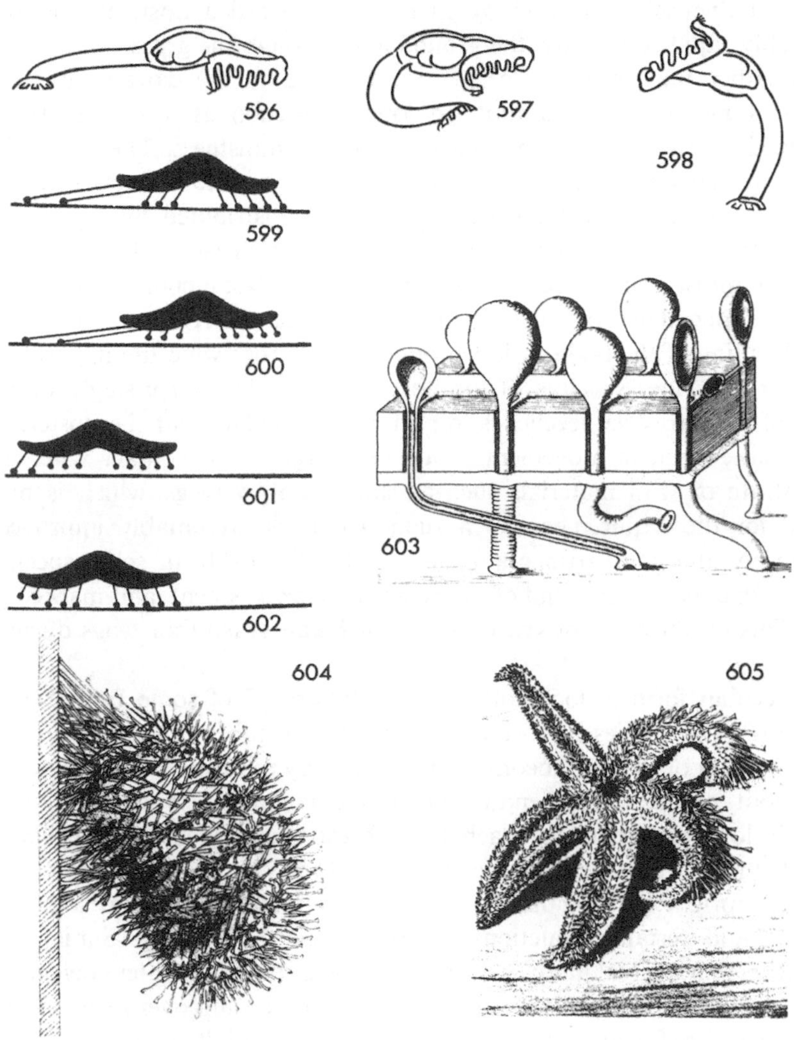

Plate 48 (Figs. 596–605)

596–598. Jump of Loxosoma saltans

599–602. Rough diagram of one kind of podia coordination in a sea star

603. Various positions of sea-star podia

604. A sea urchin climbing

605. A sea star turning over

can also orient themselves upwind by chemotaxis toward a host, proceeding at a remarkably rapid rate across the ground and through the grass.

In dry weather the terrestrial leeches are exposed to the danger that the adhesive disc may dry out and thus no longer function as a sucker. They have developed special organs which insure adequate moistness. The first pair of nephridia in many Haemadipsidae opens via long excretory ducts directly into the anterior sucker. The fluid which emerges is distributed by capillarity over a membranous corrugation of the whole sucker surface. The posterior sucker is kept moist in a more indirect fashion: the last nephridium opens beneath lobate appendages of the 23rd to 25th somites, the segmental auricles. Due to the bending of the body in locomotion, these lobes wipe the nephridial excretion over the inner surfaces of the posterior disc. The surprisingly rapid locomotion of the terrestrial leeches is in part due to the fact that the posterior sucker can make circular movements with its margin. The contact with an uneven substrate or with materials such as small knobby twigs, which is not always ideal for the requirements of a suction cup, is presumably improved by the flexibility and compartmentalization of the disc (539). In some species of Haemadipsidae the anterior end of the posterior sucker is bent, and markedly drawn out. This produces a hooked process which can grasp thin twigs during locomotion.

A most peculiar form of locomotion is the "jumping" of some Entoprocta (= Kamptozoa). The species Loxosoma saltans lives in tubes in sandy beaches under about a meter of water. It locomotes (596–598) by releasing the attachment disc at the foot of the stalk, bending down the tentacular crown until the two especially long oral tentacles reach the substrate, and then shooting itself forward with these. At the end of the jump the foot disc reattaches and the animal stands upright once again. The foot disc sometimes contains a pedal cement gland, and certainly functions at least in part by gluing. Similar in principle, although not a real jump, is the locomotion of the fresh-water coelenterate Chlorohydra. After release of the stalk it makes a somersault over its tentacles and then attaches itself again when the foot disc has come full circle.

Blepharocerid larvae can use their six suckers to move for meters over the stones in brooks, where they graze on the coating of diatoms. They move like inchworms, by alternately releasing and attaching the individual suckers, or by snakelike lateral twisting, during which they alternately lift and reattach the three anterior and three posterior suckers.

Sea urchins and sea stars travel considerable distances with their numerous podia. The podia can be coordinated as indicated in the rough diagrams of Figs. 599 to 602. As soon as the outstretched distal podia of one arm have attached, all the others detach. The attached podia shorten and pull the body along. Then all the other podia take hold; those which have done the pulling are lifted and extended anew, and the sequence is repeated. But this mode of locomotion is by no means the only possible one; in fact, it only appears when pulling is indispensable. On flat ground it is the hindmost podia (with respect to the direction of motion) which push the body forward.

The movement of the many podia is usually strictly coordinated, under control of the ectoneural nerve ring. Probably there is a subordinate control center for each arm, which sends its impulses out along the arm via the ectoneural radial nerve. It may also be assumed that mechanical stimulation of the individual podia is sensed and involved in proprioceptive control. The different arms, when they lie in various positions relative to the substrate, can show varied kinds of behavior, but the overall operation is always coordinated.

Extension of a podium is brought about by contraction of the ampullar wall, which forces water from the ampulla into the podium (603). Directional movements are produced by asymmetrical concentration of the longitudinal musculature of the podia. During locomotion the podia attach, as mentioned above, only when the mechanical situation demands it—for example, during walking on a slope and especially in climbing and turning over (605). Climbing forms like Asterias glacialis have developed extremely strong suckers. Otherwise, the sea stars walk, as it were, on a system of stilts with moderately developed suckers which are simply placed on the ground. In sand and mud dwellers (Astropecten species) the podia bear no suckers at all but rather end in conical anchoring tips.

The sea urchins climb with their podia in a very similar manner (604); their velocity amounts to about 2 cm/min. On flat ground, however, they move primarily by means of their walking spines.

Cephalopods of the genus Octopus, which inhabit rocky shores, use the suckers on their arms in a well coordinated manner, particularly during climbing. They also use them for building up circular walls of stones and clamshells, within which they live.

The water beetle Dytiscus can climb vertical aquarium walls with its tarsal suckers, even in the air. Their natural aquatic habitat, however, is rarely surrounded by such smooth banks, and the suckers may be less useful there.

2.2.6.6 Suction Mechanisms for Particle and Fluid Intake

If a small leak can be kept open between the edge of a sucker and the substrate, the pressure difference is quickly compensated by influx of the outside medium. Particles lying on the surface of the substrate are swept in by the current. If, further, a pump mechanism is available to maintain a small pressure difference in spite of the influx, this particle transport can go on continuously. This is of course the principle of the vacuum cleaner. The biological suction mechanisms considered here, which attach to a substrate in a sucker-like manner (we exclude those which operate freely in the medium), function not continuously but rather intermittently. Suction currents are produced by distension of an interior space; once this space is maximally expanded it can only be contracted. The medium which has been "vacuumed up" is then expelled through the same or another opening, and particles may be filtered out. Of course, the cavity can then be re-expanded and a cyclic filtering process carried on.

The feeding mechanism of predatory tadpoles offers an example. The larva of the toad Scutiger has a sucker (606) with marginal teeth, elongated swellings, and papillae for close attachment to the prey; at its base, there is a strong horny beak. The rasping action of the latter loosens pieces of skin, which are sucked in. A technological analog is the beater-type vacuum cleaner. Here, protrusions on a cylinder rotating within the vacuum housing beat particles of dirt out of the carpet, and these are then sucked up. Other tadpoles, of the "funnel mouth" type (607–610), hang from the surface of the water by the oral region, which is expanded into a broad plate, and suck in surface detritus. They can also attach themselves to stones with this device.

A rather specialized manner of feeding has been developed by the funnel-mouth larvae of the tree frog Microhyla. Accoding to Herter, "the lower lip is developed as a retractile funnel with ridge-like papillae (611, 612); the larva, poised at an angle of about 20° to the horizontal in the water, applies the funnel from below to the surface film of the water and spreads it out, opening its mouth (613, 614). Thus surface currents are produced which sweep the organic detritus in the vicinity toward the mouth. From time to time the lower lip is contracted, the opening between the lips is narrowed, and the food is sucked in (615)."

A tubular sucking mouth is found in many copepods of the suborder Caliigoida.

In the remaining examples of suction intake to be considered the feeding proceeds within the suction cavity, which is sealed tightly against penetration of the external medium. The gnathobdelloid leeches, which include the medicinal leech Hirudo medicinalis, have a strongly muscularized anterior suction disc. In the center are located three jaws, each bearing a hard tooth (616), which saw star-shaped wounds in the skin of the host animal. The blood of the host is drawn by pumping movements of the pharyngeal musculature. The suction developed by this leech would (in a water barometer) sustain a column of water 1 m high (for comparison, the suction generated by a pipe-smoking human is equivalent to a 3.3-m column of water). The Rhynchobdellidae also feed by suction. Some representatives of this suborder of leeches have a powerful anterior attachment disc; within this is a central pocket from which a strong, pointed sucking proboscis can be extended and stabbed into the skin of the victim. A typical example is the fish leech Piscicola geometra (617), which attaches itself to underwater plants with a similarly built posterior attachment disc and waits until a fish swims past. The discussion on page 152 applies to the physiology of this attachment mechanism.

The cyclostomes, eel-like predatory fish, include the lampreys (Petromyzontidae; 618) and the hagfish (Myxinidae). Seen from the side, the head of such a fish seems to be flattened at an angle (619), forming a powerful sucker or sucking mouth with a complicated arrangement of razor-sharp horny teeth

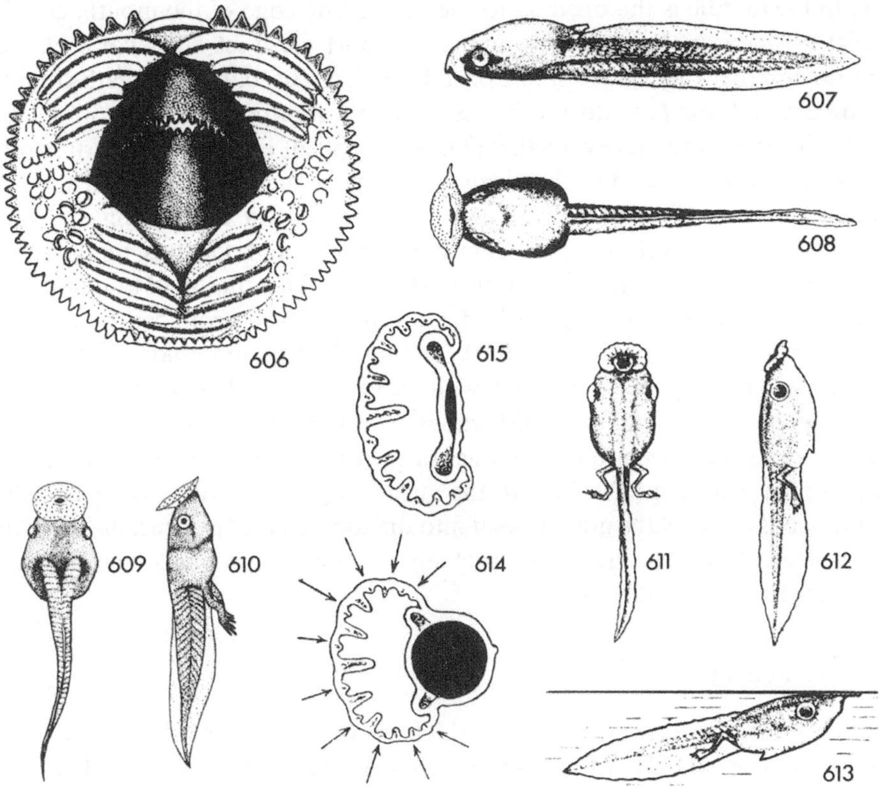

Plate 49 (Figs. 606–615)

606. Horny beak in the suction area of the Scutiger larva

607 and 608. Funnel-mouth tadpole of Megophrys montanus

609 and 610. Funnel-mouth tadpole of Phyllomedusa fimbriata

611–613. Funnel-mouth tadpole of Microhyla achatina

614. Dorsal view of the funnel-type mouth of Microhyla achatina, maximally opened

615. Funnel-type mouth of (614), fully closed

(620). In the middle is the opening to the gullet. The edge of the mouth, circular or oval in shape, is bordered by numerous short close-set cilia like eyelashes, or in other genera by larger skin appendages like the barbels of catfish. With this apparatus these predatory fishes can make a tightly sealed suction contact with their prey. The horny teeth (620) are applied closely to the substrate; the centrally placed, retractable tongue serves as a suction piston and the wall of the mouth forms the cylinder. These animals feed on both living and dead fish. The teeth rasp off shreds of skin and muscle, and the bits of tissue are sucked in together with the blood of the host. The sucking mouth of Euodontomyzon danfordi is displaced diagonally downward and is surrounded by a sturdy sealing rim with feathery lobes *l* (621). The murderous odontoidal plates $o_{1,2,3}$ as well as a rasping plate covered with little teeth *h* all operate within the suction cavity. These enable the fish rapidly to work into a prey animal. Some parasitic nematodes penetrate the tissues of their hosts in a quite similar manner. Ancylostoma caninum (622) lives in the gut of dogs. This worm grasps a villus of the mucous lining of the gut, draws it into the toothed oral capsule, disintegrates it and sucks it in. The action of sucking simultaneously pulls the animal a bit further forward in the gut.

2.2.6.7 Suction Tubules

Highly effective suction organs, in the form of funnels or tubules, are found in some protozoans. Their ultimate function is the uptake of food. These are included here as a category of suction connections because they are also capable of first producing a firm suction attachment to the (usually larger) prey.

The ciliate Didinium nasutum (suborder Gymnostomata) possesses a tubular or proboscoid protuberance of its anterior end (623). It eats paramecia by drawing them into the gullet with preformed strands of protoplasm, and from then on evidently sucks them in; the gullet is expandable and reinforced with a corona of supporting elements. Once the attachment to the prey has been made, the suction developed by these minute, barrel-shaped predators must be considerable; the hunters can swallow paramecia much larger than themselves (624) without difficulty. When several Didinium are sucking on a single prey animal, they either tear it to pieces or the strongest individual draws out of the weaker ones the parts they have already sucked in. The manner in which these forces arise is not yet understood.

The Suctoria, a second large group of ciliates, are distinguished by their tentacles or suction tubules. The cytoplasmic wall of the tentacle, which surrounds a central suction canal, can be contracted longitudinally so as to shorten the tentacle. The delicate outer wall, a projection of the pellicle enclosing the body, is thereby thrown into folds (625). Parts of prey animals are sucked in through the tentacles. There are also purely prehensile tentacles, which only grasp the prey and bring it to the suction tentacles. The genus Choanophrya and the

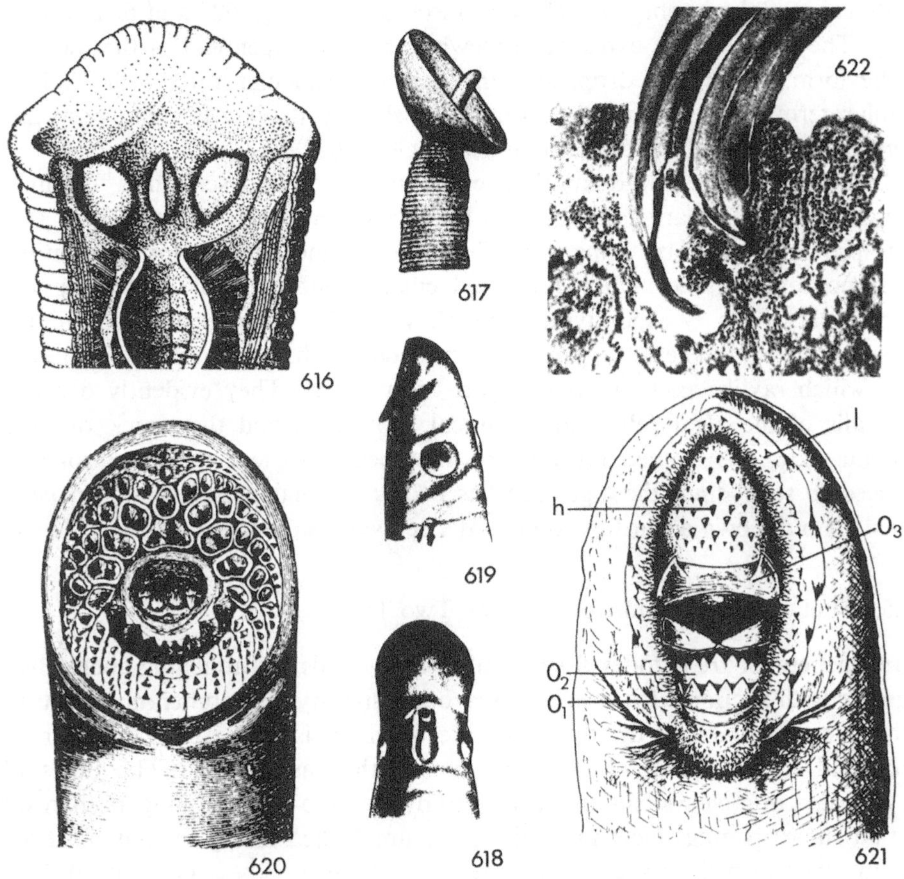

Plate 50 (Figs. 616–622)

616. Oral region of a gnathobdellid leech (Hirudo medicinalis), showing jaws
617. Oral region of a rhynchobdellid leech (Piscicola geometra), showing suction proboscis
618. Suctorial mouth of Lampetra fluviatilis, from above
619. Suctorial mouth of (618), from the side
620. Suctorial mouth of Petromyzon marinus
621. Suctorial mouth of Euodontomyzon danfordi
622. Ancylostoma caninum in the gut mucosa of the dog

ciliate-eating Sphaerophrya are shown as examples in Figs. 625 and 626, respectively. The cytoplasm is extruded somewhat from the open end of the tentacle in the form of a sticky drop, and the delicate pellicular tube with its fine, annular terminal thickening is opened out slightly (627). After the prey has been sucked in the edges of the tube nearly close (628).

In Dendrocometes paradoxus the suction tubules can be described as "terminal prongs" (630) which are gathered into bundles or branches; these in turn join more proximally to form arms (629). Such suction mechanisms for food intake occur not only in the highly developed ciliates, but even in the Rhizopoda. Vampyrella vorax, V. spirogyrae, and V. variabilis are parasitic on filiform green algae (631). These are shell-less amoebas with nearly spherical bodies from which raylike pseudopodia project on all sides. They evidently dissolve the walls of the algal cells with chemical substances and then suck out the contents (632, 633). When the amoeba first begins to feed there is probably no need to suck, since the algal cell has excess internal pressure—the plasma flows out of its own accord as soon as the cell wall is damaged.

2.2.6.8 Suction Mechanisms for Aligning Two Tubes

A few genera of trematodes form double individuals by attaching themselves together with suction cups which lock around snap-type protuberances; details were given on p. 39. At the tip of the protuberance is the opening of Laurer's canal and in the center of the cup, that of the vas deferens. The tubes of the two reproductive systems are positioned by the snap-type coupling mechanism so that their openings meet precisely. They are locked firmly in that position, and an undisturbed passage of the sperm to the egg cells is guaranteed. The connection is maintained for the duration of the animals' lives.

Temporary suction connections are found between the teats of mammals and the mouths of their young; often the teats have geometrical forms adapted to those of the palate and jaws, and vice versa. It is vital that these suction connections fit together closely, since the smallest gap would render the suction

Plate 51 (Figs. 623–633) ▶

623. Didinium nasutum
624. Four Didinium on a Paramecium
625. Choanophrya infundibulifera sucks in bits of a disintegrated Cyclops
626. Sphaerophrya with six captured ciliates
627 and 628. End of a tentacle of Tokophrya quadripartita, before and after sucking in food
629. Dendrocometes paradoxus
630. Terminal portion of the suction tubule of Dendrocometes paradoxus
631. Vampyrella vorax on a green alga
632 and 633. Vampyrella variabilis sucking out an algal cell

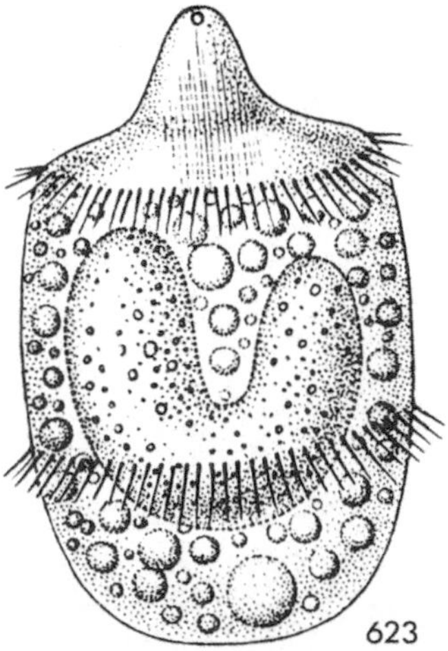

623

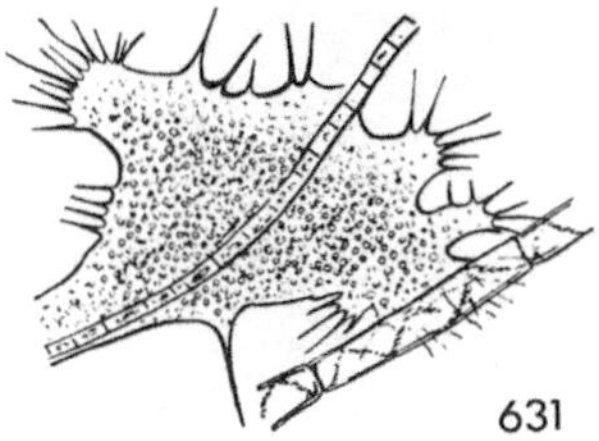

631

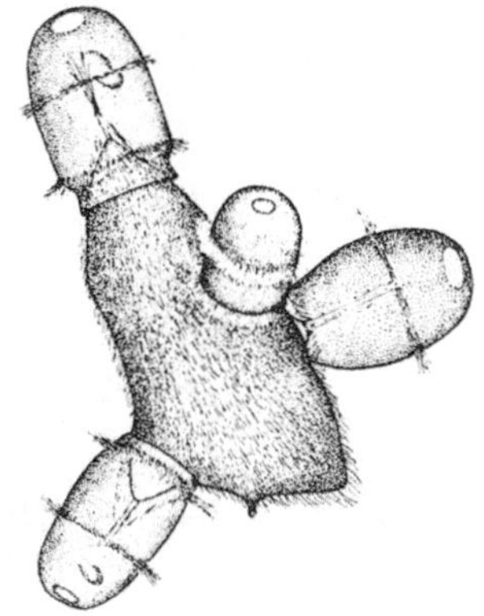

624

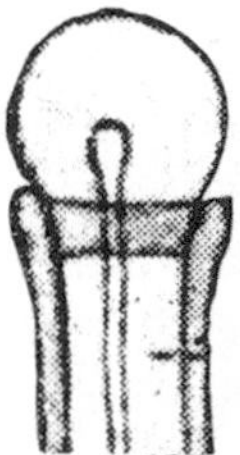

627

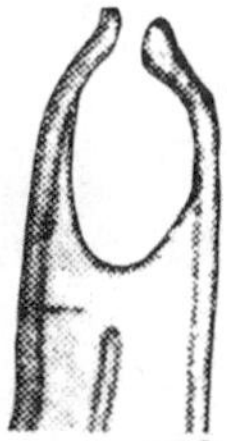

628

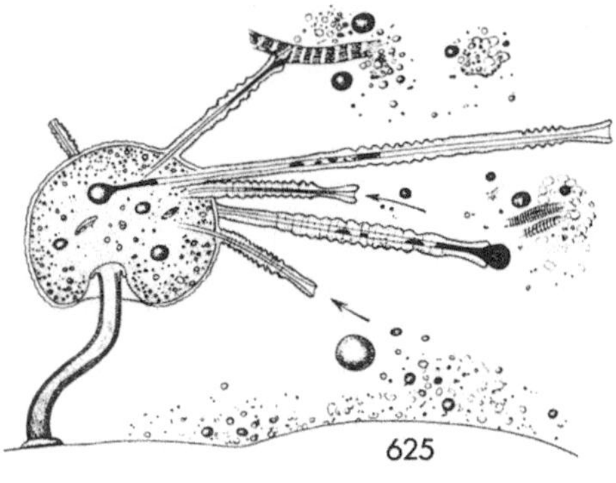

625

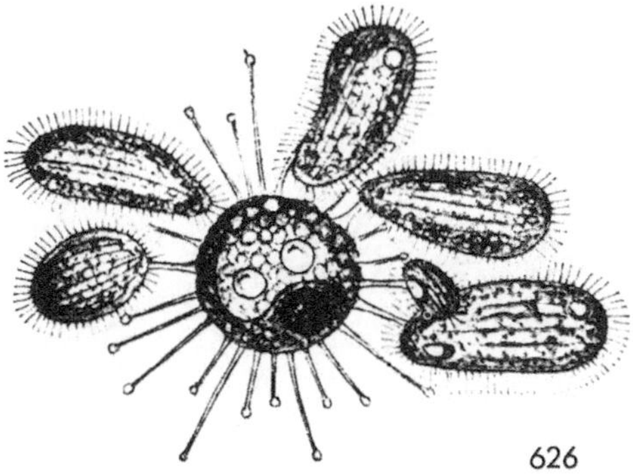

626

629

632

633

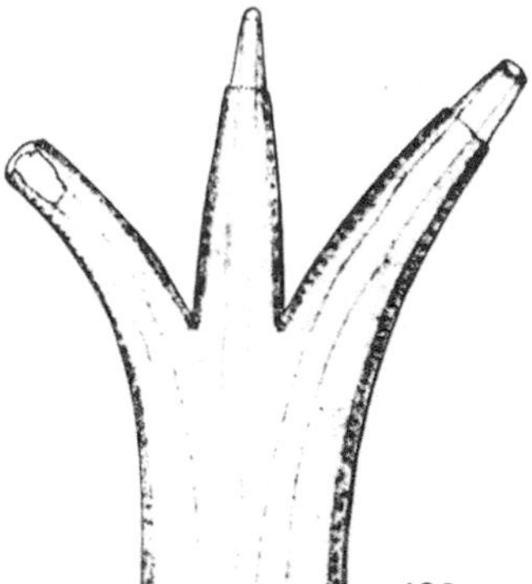

630

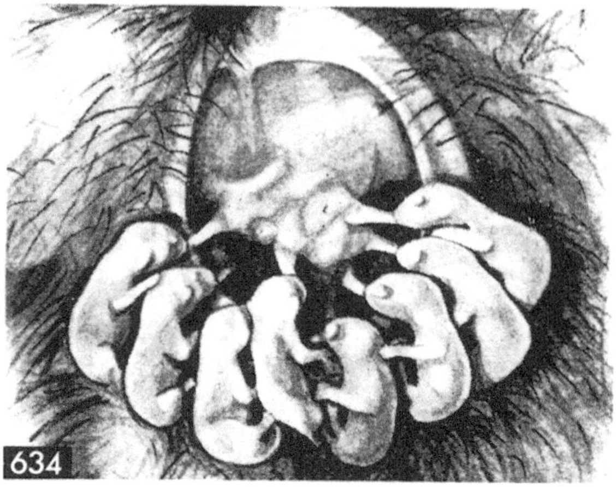

Plate 52 (Fig. 634)

Teat apparatus of Didelphys marsupialis

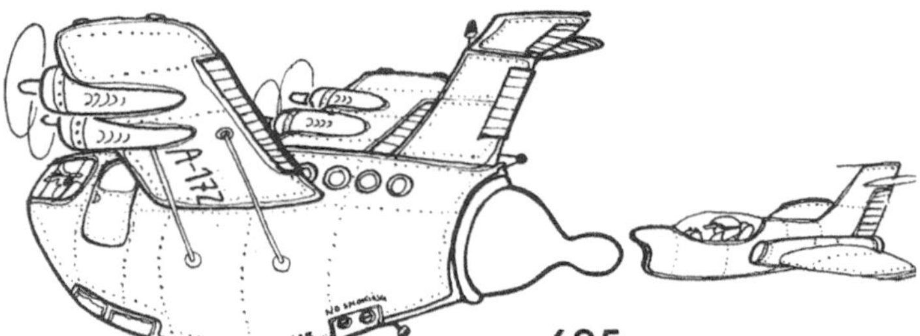

635

Plate 53 (Fig. 635)

"Pacifier" system of advanced engineering design, for in-flight refueling of fighter planes

mechanism useless. It is not surprising, then, that even the primitive mammals have developed mechanisms of remarkable design. One example is the opossum Didelphys marsupialis, the teat apparatus of which permits up to eight offspring to suck simultaneously (634).

Recently there has been divulged (presumably "leaked" through high-level channels at the Pentagon) an advanced design for in-flight refueling of fighter planes (635). This drawing makes it clear that even military technology is beginning to recognize the advantages of the pacifier principle.

2.2.7 Attachment by Glue

2.2.7.1 Basic Mechanism of Gluing

Technically, two objects joined by glue adhere to one another as a result of two different physical forces—adhesion and cohesion (636).

The glue is usually first spread on each of the two objects (A and B) to be joined; it holds to these by *adhesion*—that is, by intermolecular forces between the two dissimilar materials. For good adhesion, the glue must wet thoroughly the surfaces of the object to which it is applied. It is often difficult to effect adequate wetting; for example, a water-based glue will not stick to a hydrophobic surface. The physical interface between the glue and the object must be such that the glue sticks and anchors itself to the entire surface to be treated. It is especially advantageous if the surface is rough, with small depressions which the glue can enter by capillary action; after it has dried, it is anchored there mechanically.

If the two objects so treated are pressed together before the glue has set, the two coatings of glue combine. Now the *cohesion* (the intermolecular forces or constraints among identical molecules) of the glue itself becomes critical in holding the objects together. With a good glue, adhesion and cohesion are about as strong as the internal cohesion of the objects to be joined. If the adhesion is too weak, the bond breaks cleanly at the interface between glue and object. On the other hand, if the cohesion is too weak, the bond breaks within the layer of glue. Once the glue has set, its mechanical rigidity should be approximately the same as that of the objects to be joined. Since such objects vary from case to case, there can be no universal glue. Rather, a variety of cementing materials are used, including proteins, starches, cellulose, natural resins and waxes, synthetic plastics and resins, plant juices, natural and synthetic rubber, and other substances.

Junctions made by this technical procedure—the pressing together of two similarly treated objects—occur only rarely in nature. One example is the spawn of a genus of tardigrades (637); the smooth-walled eggs, covered with a layer of glue, occasionally stick to one another where they touch. Moreover, they stick so firmly to the moss leaflets that they are not easily washed away.

In the living world, gluing substances are ordinarily secreted by glands, and then distributed over a surface. The animal is thus prepared to attach itself to some other object. But this second object has usually not been pretreated in the same way—the stone to which an animal attempts attachment bears no coating of glue. Hence everything depends on adequate *adhesion* at the interface between the glue and substrate. Very little is known about the molecular mechanisms involved in this kind of attachment.

In many instances—for example, with wax-coated leaves and mucus-covered stones—it would seem difficult to achieve the required degree of adhesion. On the other hand, the necessary cohesion of the glue is readily attained in nature. This is demonstrated by the tensile strength of such secretions as spider webs and byssus threads; moreover, when attached animals are forcibly detached, the break is usually at the interface between glue and substrate.

Sometimes the typical gluing organs, in the form of large surfaces or adhesive discs, are lacking: spiders squeeze their threads directly out of the silk glands as quickly-hardening strands of secretion. The adhesion between the silk and both the tip of the spinneret and the inner wall of the secretory ducts is sufficient to bear the weight of the spider as it spins down the strand. Sticky mucoid glandular secretions make up the trails left by snails and the formed objects, such as rings, tubes, plates, and other shaped elements, of Thecamoeba, polychaete worms, rotifers, and insect larvae. Occasionally formed elements such as granules of aragonite (produced by the animal itself) or particles of sand are incorporated into a glued structure to reinforce it mechanically and save glue. The result is a two-component system comparable to brick-and-mortar masonry, plastics reinforced with glass fiber and, in a certain sense, to prestressed concrete. Animals use these methods to build cases and tubes. But they fall outside the category of glued attachment and so will not be discussed in detail.

2.2.7.2 General Surface Stickiness

The eggs of the tardigrade species just mentioned, Macrobiotus antarcticus, are spherical and smooth (637). A glue material is secreted through the shell,

Plate 54 (Figs. 636–644) ▶

636. Diagram of the process of bonding by glue wetting and bonding. Right above, a molecule of glue; adh, adhesion component; coh, cohesion component
637. Egg of Macrobiotus antarcticus
638. Collotheca (planktonic rotifer) with terminal cement glands
639. Reservoir for glue secretion at the hind end of the rotifer Rattulus bicristatus
640. Adhesive disc of Apsilus vorax
641. Cirripede (Lepas spec.), showing cement gland
642 and 643. Adhesive papillae of the larval Pterophyllum eimekei
644. Section through two adhesive papillae of larval Pterophyllum eimekei

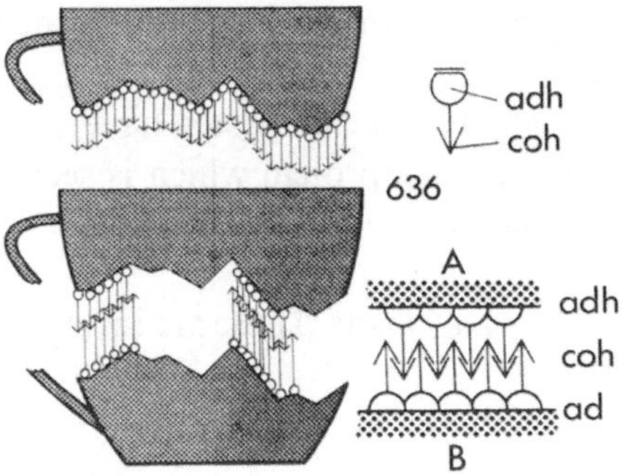

636

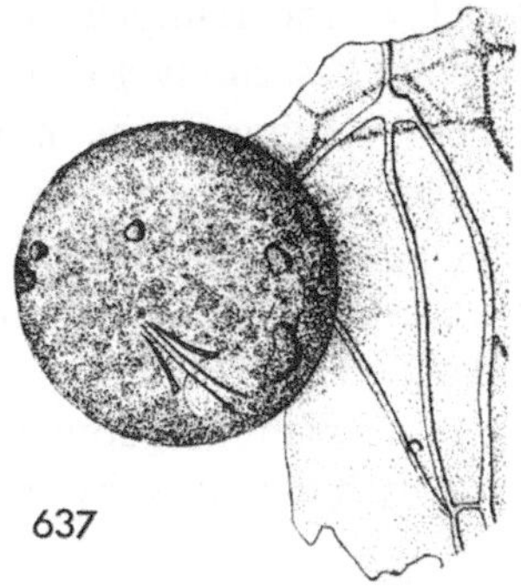
637

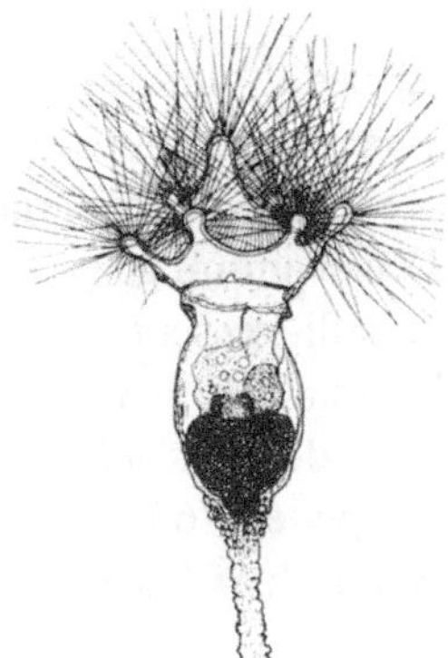
638

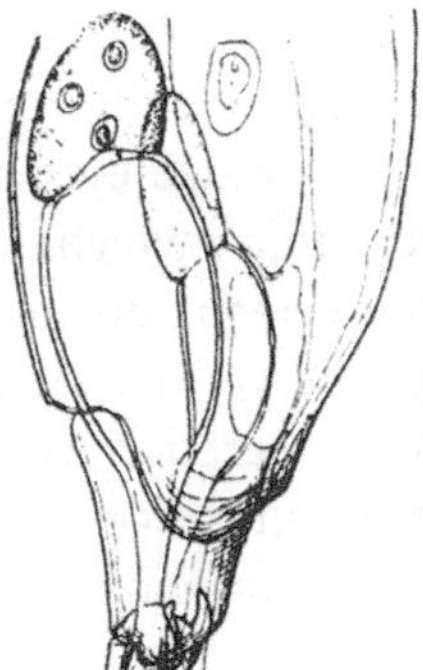
639

640

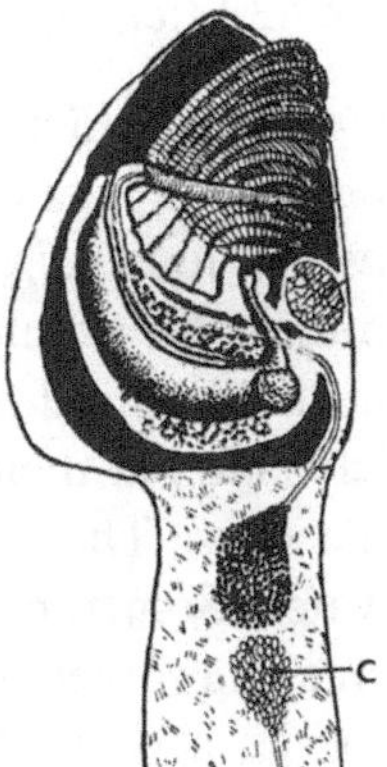

641

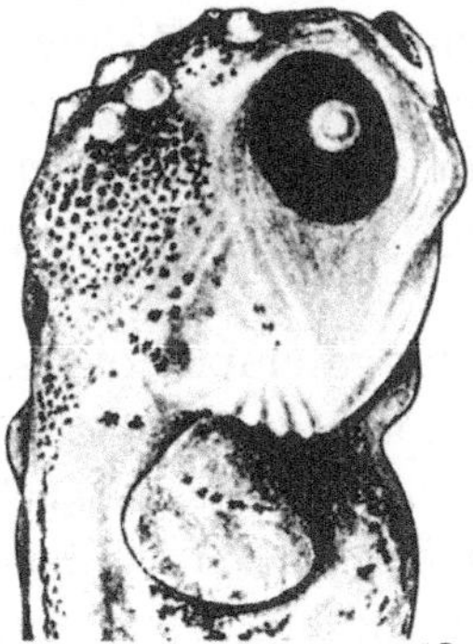
642

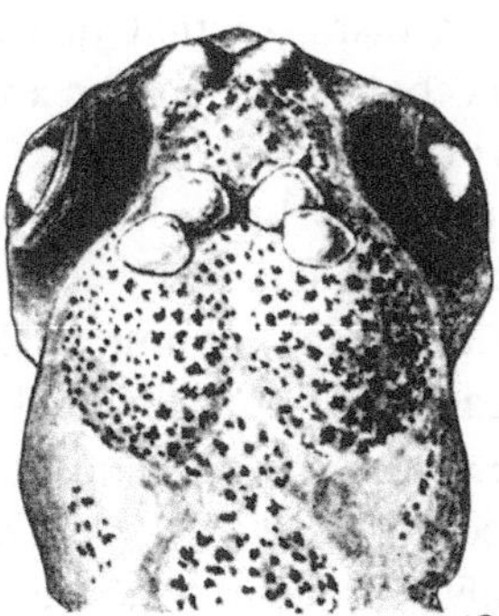
643

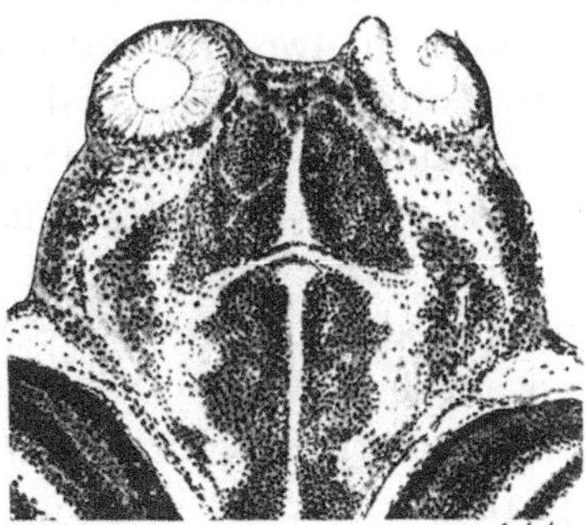
644

which is apparently finely perforated, and covers the entire surface. Thus the eggs stick securely to one another and to mosses; moreover, they gather a general coating of bits of moss and detritus.

The shells of the eggs of the fruit fly Drosophila melanogaster have a glue which cements them to plant substrates. This is a mucoprotein which is secreted by accessory glands of the female genital tract and forms a thick layer on the oocyte.

In the plant kingdom, one finds a number of sticky fruits. These are distributed epizoically—that is, by sticking to animals.

2.2.7.3 Sticky Discs

Creeping rotifers have a "foot" with two "toes". Pear-shaped pedal glands (glue or cement glands) open at the tip of each toe. In the Bdelloidea there are two or four pairs of glands, and in the Seisonidae even more. Some planktonic rotifers, for example the Collotheca species (638), which are free-swimming forms, no longer need a foot. Nevertheless, the cement glands are still present. They have altered their function and no longer serve to attach the animal but rather, the eggs. The cement glands produce a secretion which hardens rapidly in water. Sometimes it is stored in a reservoir before it is expelled (639). The creeping rotifers use the foot for leechlike locomotion, whereas in the Floscularia species it is purely for attachment.

The foot of the rotifer Apsilus vorax (Apsilidae) is reduced to an attachment disc located on the posterior part of the body (640); the animal can rotate about the axis of the disc. It is possible that the sucker principle also plays a role; especially in the Aschelminthes (= Nemathelminthes), secretory glands are often associated with suction discs, so that one cannot always distinguish between the two attachment mechanisms. This uncertainty also applies to the broad attachment discs of flatworms, e.g. the gyrodactyloid Monogenea (Trematoda) and some rhabdocoels (Turbellaria), which are richly endowed with cement glands, and to the suctorial feet of the chitons.

The barnacles (Cirripedia, Crustacea) are among the rare examples of sessile arthropods. The part of the head lying anterior to the mouth is modified into an attachment organ which may be slender and stalk-like (goose barnacles, Lepadomorpha) or broad and compressed (rock or acorn barnacles, Balanomorpha). The attachment surface always occupies the entire width of the anterior end (641). In the juvenile forms, at least, the short antennules *a* are still present. Near their ends they are swollen into a disc, on which opens a large paired cement gland *c* which secretes a glue substance. After a few days of free swimming, the barnacle larva attaches itself with this glue to rocks, driftwood, fish, ships, or crustacea. Immediately thereafter metamorphosis begins. The adult animal stays attached to the substrate for the rest of its life. This connection has great tensile strength. Even the violent action of the surf does not dislodge the barnacles attached to coastal rocks, and their attachments to the hulls of moving ships easily survive the considerable shear force of the water.

Bird mites living in wing feathers defy the strong forces of acceleration and air drag by means of hook mechanisms (see p. 58). During copulation, however, the male can no longer anchor itself in the feathers, but must hold tight to the smooth surface of the female. Males of some species solve this problem by gluing themselves temporarily to the female's chitinous exoskeleton.

The tadpoles of Hymenochirus and Xenopus have an unpaired cement gland in the region of the chin. In the first larval stages it takes the form of a backward-directed proboscis-like projection. The tadpoles glue themselves to plants and stones with the aid of this gland.

2.2.7.4 Glue-producing Papillae

In fish, too, cementing apparatus occurs. At the onset of darkness, young larvae of the cichlid Pterophyllum eimekei attach themselves to the bottom with six papillae which are characteristically arranged on the top of the head (642, 643). They do this until they are about eight days old. In the middle of each papilla there is a cavity into which the long goblet cells of the mucus gland discharge. The cavity, in turn, opens to the water through a pore (644). Tough, sticky mucus emerges in a strand through the opening. As soon as ten days after hatching, however, these organs have degenerated entirely.

Cementing organs on the top of the head are known only in teleosts (the bony fish). Besides Pterophyllum, they are found in the species Hyperopisus bebe (Mormyridae) and Sarcodaces odoe (Characinidae). In other fishes, attachment organs of this sort are located only on the underside of the head; these are known, among the non-teleost Actinopterygii, in the genera Polypterus, Lepidosteus, Amia, and Accipenser, and in certain lungfish (Dipneusti)—Protopterus and Lepidosiren.

2.2.7.5 Glue-producing Tubes

The gastrotrichs (Aschelminthes = Nemathelminthes) inhabit the floor of bodies of fresh water and the ocean. Only 0.07 to 1.5 mm long, they are among the smallest of the worms. They live in the fine interstices of the upper layers of sand. Parts of the body surface are covered with cement tubules; in one of the two orders the anterior, lateral and posterior portions are involved and in the other, only the posterior parts (645–647).

The tubules are usually cylindrical, with strongly sclerotized walls. At the end is a pore through which one or two cement glands open (648). Occasionally the tubules are reinforced internally by a somewhat flexible skeletal rod. Usually there are muscles which can aim the tubules in different directions. The details of their shape can vary greatly from species to species, and even from place to place on the body of an individual. Sometimes, among the gastrotrichs, the tubules at the end of the body are grouped together on muscular, movable skin papillae (649). In Chaetonotus the tubules open separately on two long processes, the "toes" (650), in Tetranchyroderma they are gathered into groups of three (651) and in Dactylopodalia, into groups of seven (652).

With these gluing tubules, the animals can attach, on contact, to grains of sand. By periodically attaching and releasing the front and hind ends they can also creep like a leech. Only the genus Neogossea lacks such tubules; it lives among the free fresh-water plankton, where there is no place for it to settle. In place of the tubules its hind end is provided with a "parachute" of spines which can be spread out.

Most of the kinorhynchs, another class of the Aschelminthes, have two ventral cement tubules on the anterior part of the trunk. Like the gastrotrichs, these purely marine worms are only 0.5 to 1 mm long; they live on the floor of the sea, where they burrow through the sand with the circlet of hooks on the spherical "head" (see p. 73). Finally, the hind ends of many male roundworms (Nematoda, the largest class of Aschelminthes) are drawn out into a tube, at the tip of which open the secretory ducts of numerous cement glands.

2.2.7.6 Sticky Surfaces for Prey Capture

Some genera of insectivorous plants live on high moors, on ground which is poor in nitrogen. They obtain a high proportion of the nitrogen compounds they require for life by catching insects on sticky projections. They then digest their captives, except for the chitinous exoskeleton, by means of secreted proteases and other enzymes. In the butterwort Pinguicula (Lentibulariaceae, Tubiflorae), the 25 species of which are distributed in the northern temperate zone, one finds both stalked and unstalked glands on the surface of each leaf. These arise from epidermal cells. The stalked glands produce a sticky secretion for capturing prey and the others, a digestive secretion. Another insectivorous plant family is the Droseraceae (Sarraceniales); the genus Drosophyllum (one species, in Portugal) and the sundew Drosera (eighty circumpolar species) have appendages in the form of long sticky tentacles (653), with both cement and digestive glands. In the course of a few hours after capture of an animal, they wind themselves actively around it (654). Similar techniques of prey capture are found in the Australian genus Byblis (Byblidaceae, Rosales), and even in a fungus, the phycomycete Zoophagus insidians. The latter lives on moist ground and catches rotifers in its sticky mycelium.

Among the animals which catch insects in a similar manner, the honeytail Dulcicauda griseaurella (438), mentioned previously, deserves further attention. As the generic name implies, epidermal glands on the distal part of the tail produce a sweet secretion; this attracts insects and, because of its sticky consistency, holds them fast. Once an insect has been caught, the tail relaxes. This contrivance, as simple as it is effective, brings the prey within range of the

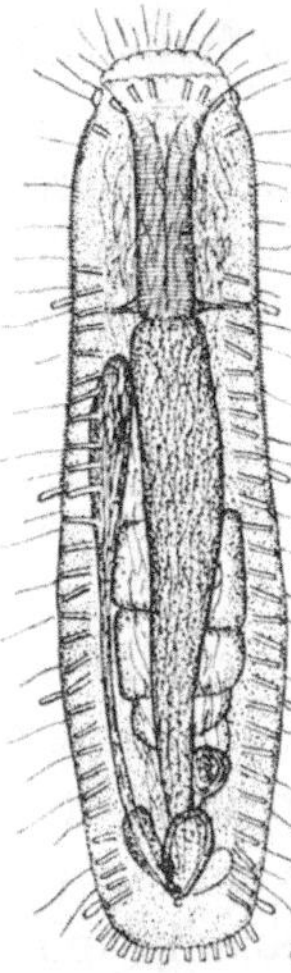

645

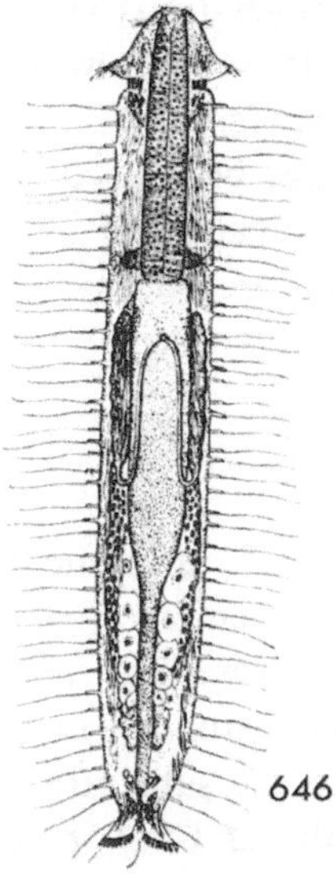

646

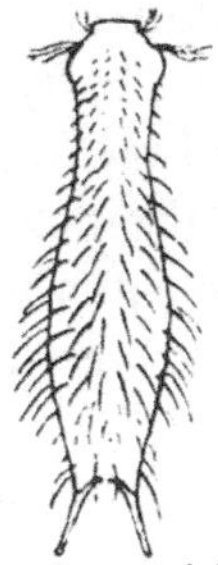

647

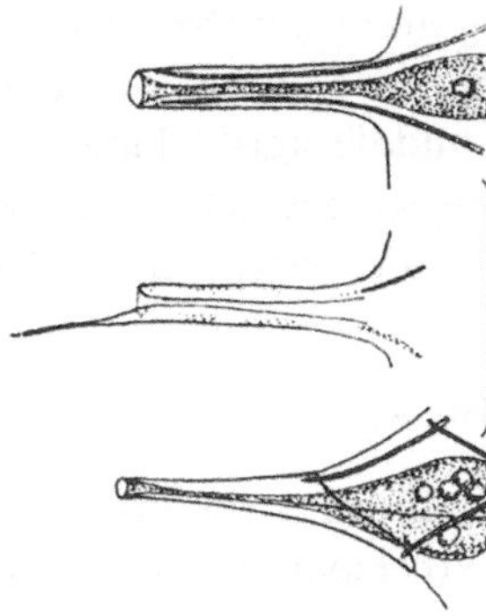

648

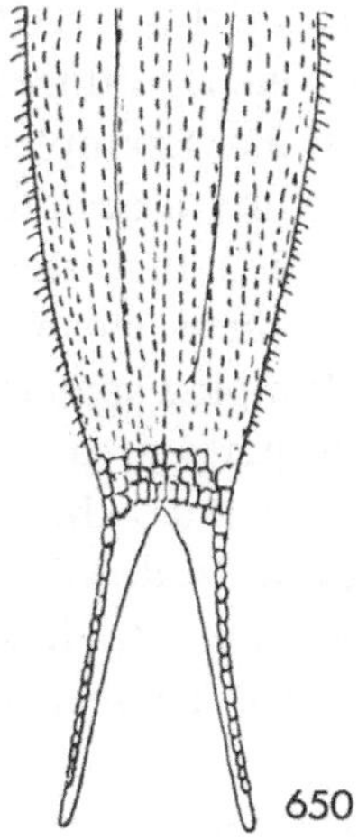

650

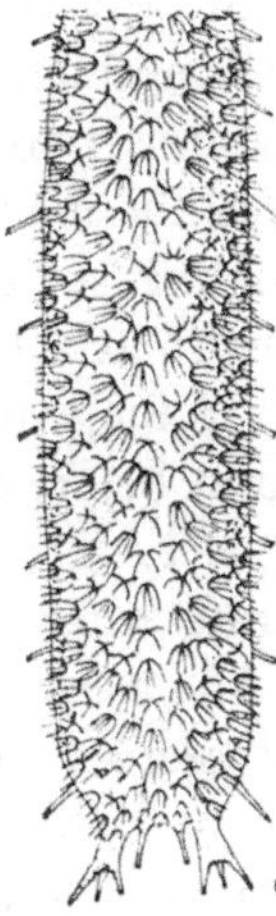

651

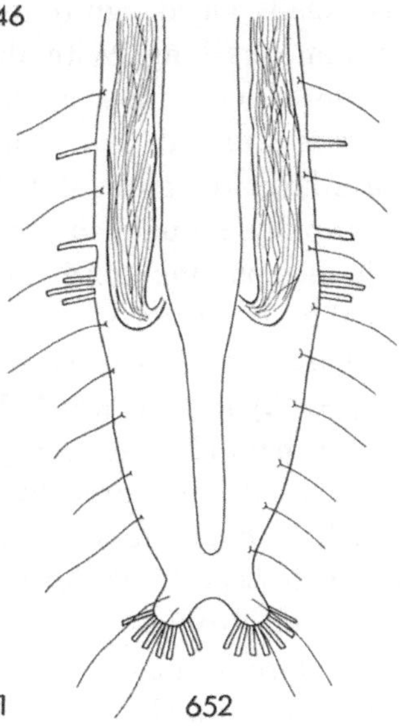

652

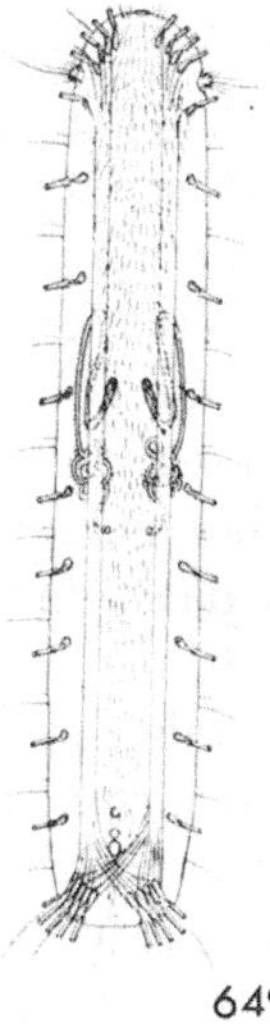

649

Plate 55 (Figs. 645–652)

645. Glue tubules in a gastrotrich
646. Glue tubules in a gastrotrich
647. Glue tubules in Chaetonotus spec. (Gastrotricha)
648. Opening of the adhesive glands in the tubules of gastrotrichs
649. Diagram of the muscles which position the caudal glue tubules of gastrotrichs
650. Groups of glue tubules in Chaetonotus
651. Groups of glue tubules in Tetranchyroderma
652. Groups of glue tubules in Dactylopodalia

forelimbs which pluck it out of the trap. Reports (thus far unconfirmed) have it that in the ritual feasts of the Polynesians there are always a few honeytails standing about on the table to serve as flycatchers.

Man-made objects of identical function have been used since ancient times. In the middle ages, "lime-twigs" (branches spread with a sort of glue called bird-lime) were widely employed to catch songbirds. And before the introduction of insecticides, streamers of flypaper were a common sight.

2.2.7.7 Sticky Threads

A mechanism for the production of sticky threads is implicit in the previously described cement papillae of the fish Pterophyllum eimekei (642–644). The cement glands discharge their secretion into a central reservoir, and from there it is squeezed out through a nozzle to form a fine strand. The fish, which in the first few days of life is usually glued tightly to the substrate, occasionally spins out threads several centimeters long. With the ends of these attached to the ground, the fish is anchored as it rides in the current. The process is rather like extruding strands from a tube of glue—in both cases the material is forced through a nozzle from a large reservoir. But if the mechanism which produces the secretion operates rapidly enough, the reservoir is unnecessary. Examples of this situation include the spinning nozzle used in the manufacture of acetate thread for fabric, and the spinnerets of web-building spiders. As a first step in making a web, the spider Aranea diadema constructs a bridge between two branches or stalks, using sticky threads which float in the wind. Jumping spiders (Salticidae, Araneae) glue themselves to the ground before they jump at their prey; during the jump they are anchored by a safety line strung out behind them. Other spiders stick a thread to the substrate when they walk down a vertical surface, or simply descend by the thread. The webs which some spiders spin are provided with little droplets of glue at regular intervals. These act like lime-twigs, for the capture of prey.

Even underwater animals use sticky nets as traps. The aquatic snail Vermetes gigas turns its vermiform, twisted shell so that the opening faces upward. When

Plate 56 (Figs. 653–662) ▶

653. Adhesive tentacle of Drosera
654. Movement of the tentacles on a leaf of Drosera rotundifolia
655. Byssus threads of Mytilus
656. Thread-producing glands in a marine nematode, Thoracoston
657. Discharge of the cement secretion in an enoplid nematode
658. Pupa of Liponeura
659 and 660. An example of the Symphyla, dorsal and lateral views
661. Use of larva as a "shuttle" by Oecophylla smaragdina
662. Oecophylla smaragdina weaving leaves together

653

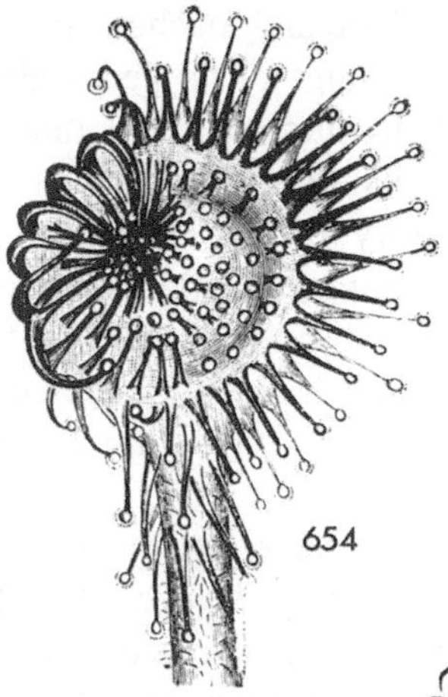

654

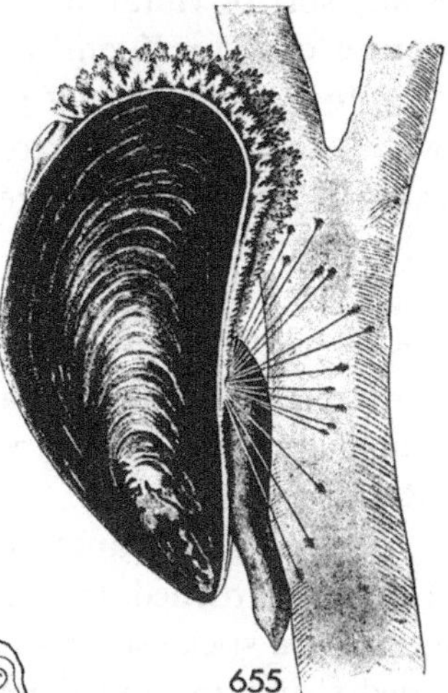

655

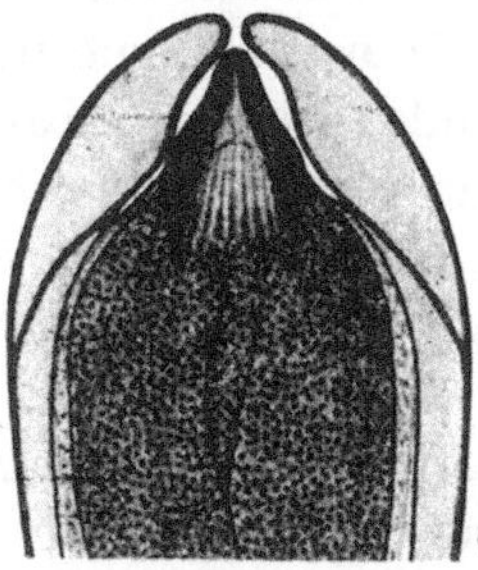

656

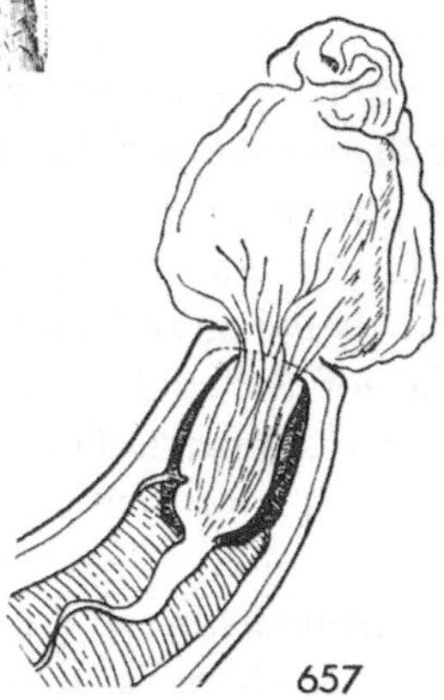

657

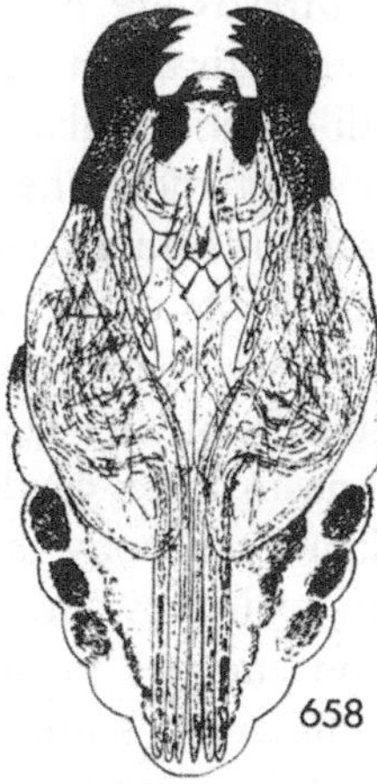

658

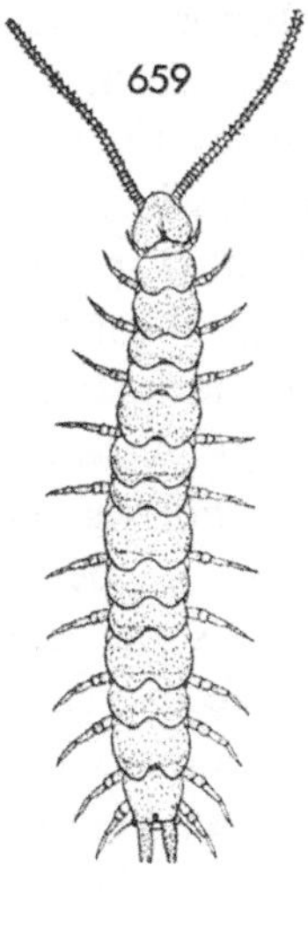

659

660

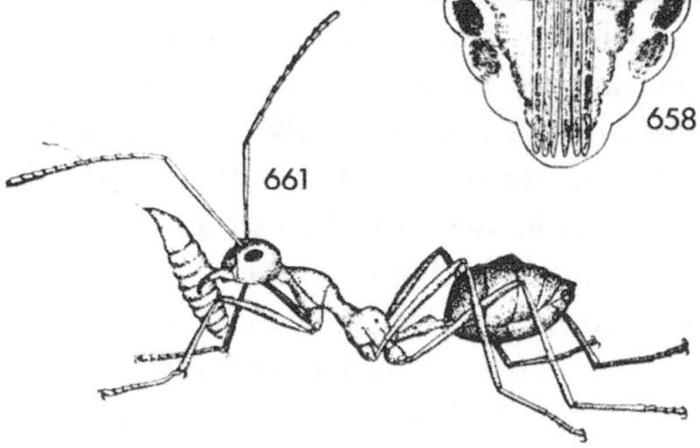

661

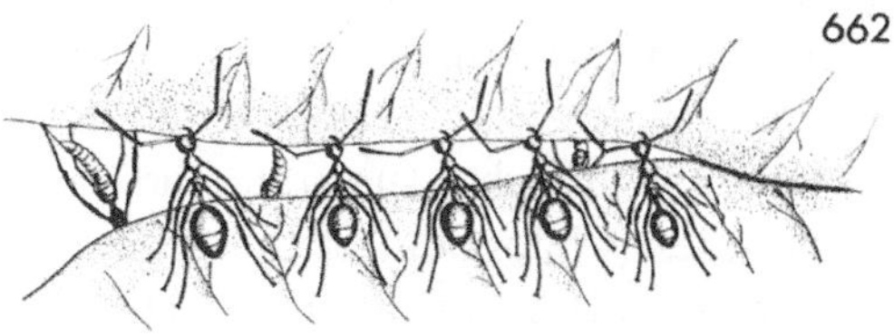

662

it senses any sort of vibration, even that from the tiny movements of Daphnia at a distance of 5 to 20 cm, the snail suddenly pushes three or four mucus threads, up to 30 cm long, out of the openings of its pedal glands. These float in the water, and small plankton stick to them. After a time the snail draws the threads, together with the prey, into its mouth.

Attachment to the substrate by sticky threads is by no means confined to the spiders, but is widespread in the animal kingdom. Rotifers, as we know, have both cement glands and attachment discs at the foot (638). Some species draw their secretion out into a long thread, which is occasionally dragged along when the animal swims. This is frequently the case, for example, in the Trichocercidae (= Rattulidae).

Many of the lamellibranchs attach themselves by means of byssus threads. Once they have hardened, these horny threads have great tensile strength. They are produced as a sticky secretion of a pedal gland. The byssus is the remarkably complicated structure formed from the secretion of the pedal glands into the byssus pit, which is continous with a groove running along the foot. The externally visible threads are produced by secretion in this groove; to attach them, the animal extends the movable foot and presses the tip of the groove to the substrate. A small accessory gland surrounds the strand then extruded with an encasing layer and provides the material for the sticky "attachment plate" with which the strand is firmly fastened to the substrate. One thread after another is sent out. The edible mussel Mytilus (655) constructs a strong yet elastic connection to the substrate in this way.

Some free-living nematodes (roundworms), particularly the marine forms, have one to three caudal glands which open at the end of the body. In the Enoplidae, the body is drawn out into a cuticularized spike with the gland opening at its tip, surrounded by a cupola with a central pore (656, 657). The secretion emerges in fine streams and hardens in the water to threads with which the animals attach to solid bodies. In the pupae of the Blepharoceridae (Diptera) there are two or three flattened places on the ventral surface, through which fluid chitin is said to be extruded. This glues the pupa firmly to stones and plants in the upper levels of rapidly flowing water. The pupae of the widespread blepharocerid genus Liponeura (658) have attachment organs located in pairs on the fourth, fifth, and sixth abdominal segments. They cement themselves to the substrate in such a way that the broad end of the head points downstream. The Symphyla are centipede-like animals with twelve pairs of legs. A pair of spinning cerci (659, 660) is formed from the limb buds of the thirteenth segment and thus represents a modified pair of walking legs. At the tip is the opening of a silk gland which may extend as far forward in the body as the sixth segment. When they touch a substrate, the cerci can produce an effective sticky thread. Millipedes (Chilopoda) of the genera Lithobius and Polybothrus even launch sticky threads at an attacker. They produce these with single-celled epidermal glands located in great numbers on the last four legs. The fifteenth pair of legs can have as many 500 of them. When defending itself against large ants or spiders, Lithobius turns its hind end toward the enemy and beats the posterior legs rapidly up and down. The glands extrude droplets of a viscous secretion which coalesce and fly off as highly elastic sticky strings. These stick to the attacker and can glue its legs together as effectively as if they had been bound with rope.

It is common knowledge that leaf-cutting ants build large nests, usually of vegetable material; the components of the nest are glued together, edge to edge, with threads spun by the ants. This procedure involves some surprisingly sophisticated behavior. For example, the imagoes of the ant Oecophylla smaragdina lack silk glands, but the larvae of this species do have them at the anterior end. When joining two leaves some of the adults each take a larva in the mandibles (661) and use it as a "shuttle." The operation proceeds as follows: several ants line up side by side outside the nest on one leaf (662) and, holding it firmly with their legs, they pull in concert on another leaf with the mandibles, dragging it slowly into position. Then the "weavers" appear on the inside of the nest, each holding a larva in the mandibles so that the gland points forward and upward. At the point where it is held, the larva is rather strongly compressed, suggesting that the pressure may activate the silk glands. The larvae are dabbed onto the edge of a leaf, where they glue down their threads. Then they are carried diagonally over to the adjacent leaf, touched down again, and so on, back and forth. Each time they touch the larva to a leaf, the weavers pause for a moment so that the thread is firmly glued down. After an hour or so the leaves are connected, without a gap, by a finely-woven fabric. Not all leaf-cutting ants weave in this manner, and it is interesting that well-developed silk glands are found in just those larvae which are used as shuttles; in these species the glands are thick tubes which wander throughout the body (663).

2.2.7.8 Combined Glue-and-Hook Fasteners

The larvae of the blackflies or buffalo-gnats (Simuliidae) are to be found in any brook. Some species also live in rapidly flowing mountain brooks. These (664) cement themselves to stones and plants by their hind ends and extend the fanlike oral appendages into the current. They filter out small organisms, organic particles and bits of mud. The resistance this filtration apparatus offers to the current necessitates a firm anchorage on the substrate. The terminal anchor plates of these larvae consist of a coronet of two-pronged hooklets (665). But the hooks are not fastened directly to irregularities in the substrate. Rather, the larva secretes from its salivary glands a tough mass of threads which it glues to the stone. It is this mass that serves as a secure anchoring ground for the hooklets of the coronet. Similar hooklike rings occur in the larvae of Deuterophlebia; the attachment organs are formed from lateral outgrowths of the abdominal segments. These animals are found in the mountain streams of central Asia and Yellowstone Park in North America. The larva of the blackfly Simulium variegatum, using this combined method of attachment, can withstand current velocities up to 3 m/sec; such velocities sometimes do occur in raging mountain brooks (the terminal velocity in a waterfall is 6 m/sec). Ordinarily one finds these larvae only in regions where the current velocity is 1 to 2 m/sec (666, lower histogram). If one experimentally raises or lowers the current velocity and observes the distribution of detached larvae (666, upper histogram), two groups are seen; with higher current velocities they are presumably torn free passively, and with lower velocities they actively detach themselves for reasons as yet unknown. Similar behavior has been observed in other Simuliidae.

2.2.7.9 Sheets of Mucus

Spiders of the family Theridiidae use their silk glands for prey capture, but not in the manner of the net-spinning spiders already discussed. These spiders actually use the comb on the tarsi of the fourth pair of legs to fling the contents of the glands over the struggling victim, covering it with a sticky mucoid sheet which completely immobilizes the prey. The terrestrial planarian Rhynchodemus (Turbellaria, Platyhelminthes) catches fruit flies by sticking them to the substrate with a mucous layer on its uplifted anterior end. The marine slipper limpet Crepidula has a glandular ridge at the base of the gills, which secretes a broad carpet of mucus with very fine pores. Particles of plankton stick to its surface as the respiratory current sweeps by. Finally, it is rolled up like a real carpet, along with its load of food, and brought to the mouth like a sausage to be eaten. A second mucus-sheet filter, but with a larger mesh, is formed at another place. Elementary forms of this astonishing behavior are found in the gastropod genera Viviparus and Struthiolaria.

On the foot of the terrestrial snail there are thick cushions of glands. As the snail creeps along, these continuously secrete mucus, which forms a thin film between the sole and the ground. The snail leaves behind it a solid carpet of mucus which sticks to the ground and soon dries to a light, shiny band. The function of this mucus sheet, with respect to the mechanics of the process of creeping, is possibly rather complex; it has not yet been investigated.

2.2.7.10 Structures Formed from Hardened Secretions

Secretions from cement glands can solidify, by drying out or by chemical hardening, to form structured elements. The substance may be secreted into a cavity, so that it hardens into a corresponding shape. The byssus of the mussel Mytilus (667) is formed in this way. But the secretions can also be extruded centrifugally from a ring of cement glands on a cylindrical body, forming a belt or girdle of mucus. Some earthworms produce such a cylinder of secretion in the region of the clitellum during copulation (668). After it dries, the structure becomes

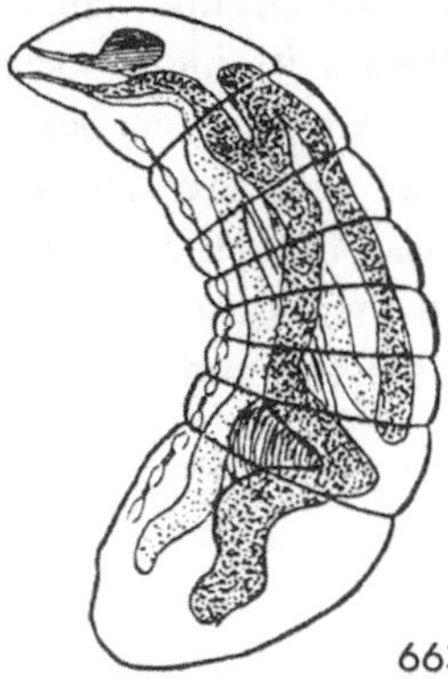

663

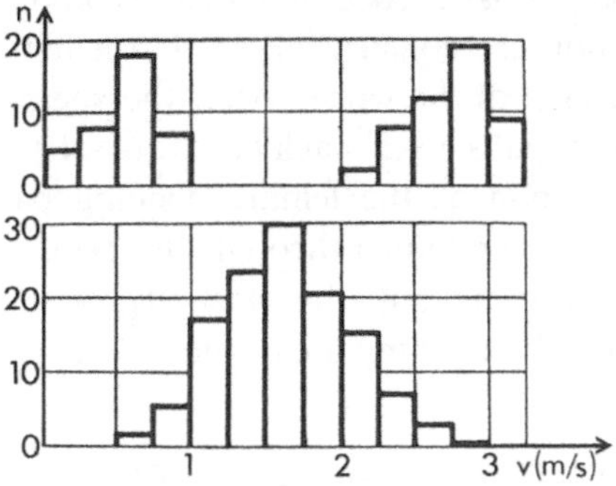

666

664

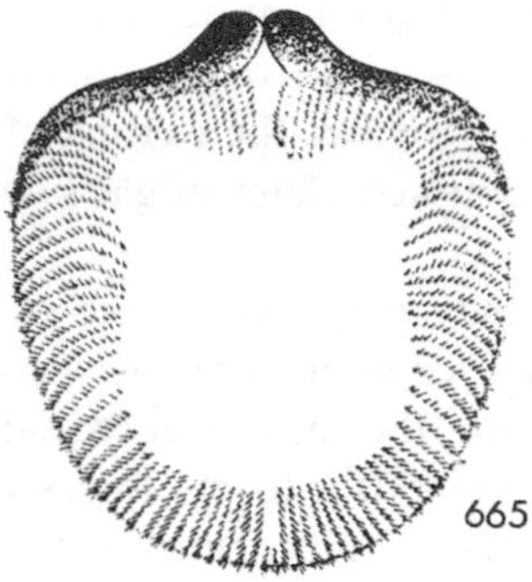

665

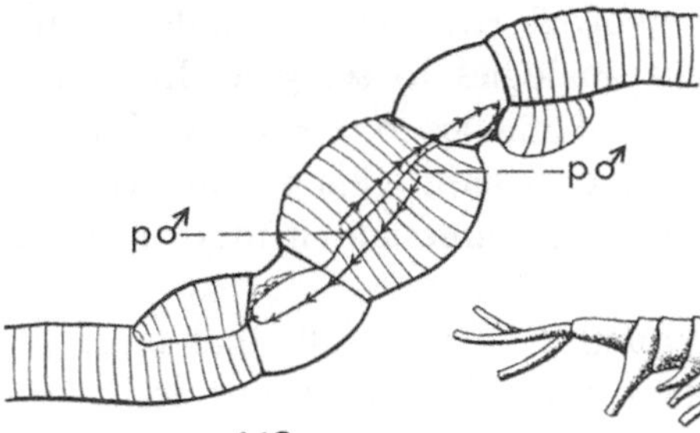

669

668

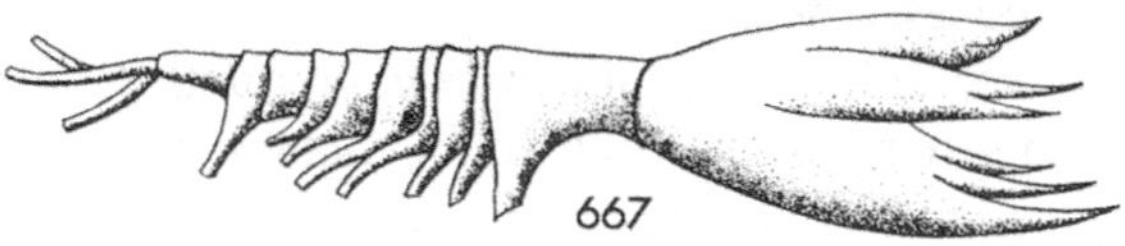

667

Plate 57 (Figs. 663–669)

663. Silk glands of the larva of Oecophylla
664. Simuliid larvae attached to a pebble in a brook
665. Anchor plate of a simuliid larva
666. Distribution of the larvae of Simulium variegatum, in terms of the numbers (n) of larvae attached (lower histogram) and free (upper histogram) at different current velocities (v)
667. Byssus of Mytilus edulis
668. Formation of secreted cylinders by lumbricid worms
669. Egg capsule of a lumbricid worm

a rubbery elastic cocoon which is under tension. The two circular ends of the cocoon are constricted, so that a lumen remains between its inner surface and the skin of the worm. After the sperm has reached the seminal receptacles, the worm pulls itself backward out of the girdle; first the eggs are deposited in the cocoon as the female opening passes through, and then the sperm is ejected into the lumen through the pore *p*♂ (668). When the worm has pulled itself completely out, the elasticity of the structure causes the ends to close, automatically forming a capsule for the eggs (669).

2.2.8 Intermolecular Adhesion

Adhesion, as the word is used in physics, is the intermolecular attraction between the closely contiguous surfaces of adjacent materials. Adhesive forces are involved in the process of writing with chalk or pencil, the clinging of water drops to the rim of a glass, the wetting of solid bodies by fluids, the sticking together of highly polished plates of glass, gluing, soldering, and the adsorption of gases on solids.

As already emphasized, the organs of attachment of animals and plants can involve combinations of suction, gluing, and elementary adhesion. For example, with the suckers of cephalopods and blepharocerid larvae glue contributes less than 10% to the forces resisting separation, whereas in the sea stars which have been studied the proportion is closer to 50%. Moreover, it is fundamentally impossible to draw a line between phenomena in terms of gluing vs. adhesion *per se*, since identical physical forces are involved. Even with the watery sealant in an "adhesion" contact there are strict analogies to set glue. In some of the biological literature, though, it has been the practice to separate such phenomena. One criterion commonly applied is to use the term "gluing" if the secretions of attachment organs are sticky or mucous, and "adhesion" if they are watery or of low cohesiveness.

This terminological problem is compounded if one recalls that most such attachments depend critically upon factors other than adhesion alone. First, glued junctions, no matter how cohesive the glue, are vastly strengthened if the surfaces to be joined are rough. In the extreme case, the hardened glue may be so well anchored in irregularities that the actual forces of adhesion are of secondary importance.

Second, even "pure" adhesion attachments—those which are instantly fully functional since the sealant need not harden at all—are physically quite complex. For example, because of the atmospheric pressure and the viscosity of the sealant in such junctions, they may withstand appreciable transient tensions which, if maintained or built up very slowly, would easily pull them apart. That is, even if a junction is not designed so as to seal tightly at the rim and thus act as a proper suction cup, a pull does result in a pressure reduction within the sealing substances. The typical "cavity" of the real suction cup, of course, may not be visually obvious, since the sealant changes volume less than does air when the pressure is reduced. The pressure differential acts for the moment, along with the adhesive forces, to counter the applied pull just as in a suction cup. The low internal pressure can be relieved, of course, by the flow of air or water into the joint along its border, but this can be a slow process, resisted by the viscosity and other properties of the sealing material. If the pull is brief, the attachment holds fast.

This transient effect can be demonstrated by laying a ruler so that half of it projects beyond the edge of a table; then either wet the ruler-to-table junction or place a piece of wet newspaper over the half on the table. Now strike the projecting end sharply. The ruler breaks because air cannot flow into the junction quickly enough to relieve the induced pressure differential between junction and atmosphere. On the other hand, somewhat smaller tensions can be tolerated by such junctions even though the pull is maintained; that is, the effects both of capillarity (forces due to differences between the strengths of cohesion in the sealant and adhesion at the interfaces) and of the adhesive forces themselves may suffice to maintain a small pressure differential or even by themselves to counter moderate stresses on the junction. For example, a wet piece of paper or a bit of foil sticks quite well to a smooth plate of glass.

Certain so-called "adhesive organs", which have been described primarily in the tree frogs and insects, will be discussed here under the traditional heading of "adhesion." But, as is quite clear from the above considerations, attachment is really due, in every case, to the combination of several physical mechanisms.

2.2.8.1 Simple Adhesive Surfaces

In some tropical tree frogs the entire skin of the belly has been modified into a large, flat adhesive disc which assists the action of the attachment organs on the fingers and toes. The adhesive disc is particularly effective when the frog lands after a jump; if the jump is to a small target, such as a leaf, the frog must stop on the spot. A tree frog can also attach itself to a vertical wall of glass with the belly skin alone—fingers and toes need not touch the wall at all.

The belly skin of the tree frogs (Hyla) is not smooth, like that of aquatic and grass frogs (Rana), but rather consists of many close-set, rounded papillae (670). On each of these, the openings of several glands release a watery secretion. Internally, the morphology is also more specialized in Hyla. Bands of connective tissue run perpendicular to the skin surface. Between these are bundles of smooth muscle, and above the M. rectus abdominis on each side there is a single layer of fibers of the M. obliquus externus. In contrast to those in Rana, these muscles are connected to the skin: little pillars of connective tissue branch off from the fascia and run to the skin of the abdomen. The pillars are bundles of connective-tissue fibrils which are continuous with those of the fascia. Hence in Hyla the M. obliqui externi and M. pectorales can exert tension, through the abdominal lymph sac, upon the abdominal skin.

When the tree frog sticks to glass, one can observe that the interpapillary grooves are flattened against the glass and filled with a watery fluid. Thus there is not the expanded central cavity one might expect if the muscles were drawing on the middle of the disc to produce the ordinary suction-sup action. However, as mentioned in the discussion above, a suction cup requires only a good seal and reduced internal pressure. The effect is demonstrated by a simple experiment*: stick a good-quality, pliant suction cup firmly onto the bottom of the inside of a glass full of water, first taking care that the cup is free of bubbles. Then dump out the water. A visual study of the stuck suction cup through the clear bottom reveals no obvious bubbles or gaps, for the moderate but crucial reduction of pressure (due to the elasticity of the cup) neither "boils" the water film nor frees appreciable gas from solution. A very strong pull on the suction cup, however, can reveal an obvious bubble. The fact that it vanishes from sight when the pull is reduced is a graphic demonstration of the extent of the vacuum generated beneath the cup.

Whether or not the tree frog uses muscularly induced suction in this way (the indicated experiments have apparently not been done), it is clear that elementary adhesion must at least play a distinct role. The small mass of the animal and the excellent fluid seal indicate that the physical effects discussed above in connection with the "piece of wet paper on glass" are relevant.

2.2.8.2 Compound Structures

Five broad adhesive discs adorn each hindleg of the tree frog, at the tips of the toes (671). On the foreleg only four are well developed. The force they sustain when attached is considerable. They easily support the weight of the body, allowing the animal to attach to smooth overhanging surfaces which are hard for its enemies to reach. Direct adhesion is certainly involved. The epidermis of each disc is bordered by very slightly keratinized cuticle, the cells

* The author would like to thank the translator for suggesting this clarifying demonstration.

Plate 58 (Figs. 670–676) ▶

670. The tree frog Hyla spec. on a glass plate
671. Adhesive discs on the ends of the toes of a tree frog
672. Epidermal cells in the adhesive pads of a tree frog
673. Cells of the "typical" epidermis of a tree frog
674. "Trumpet bristles" on the galea of Axinotarsus pulicarius
675. A trumpet bristle of Malachius bipustulatus
676. Transport of pollen by the trumpet bristles of Malachius bipustulatus

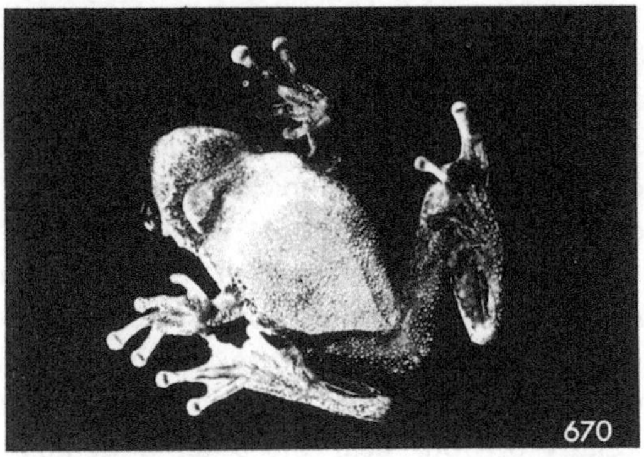
670

671

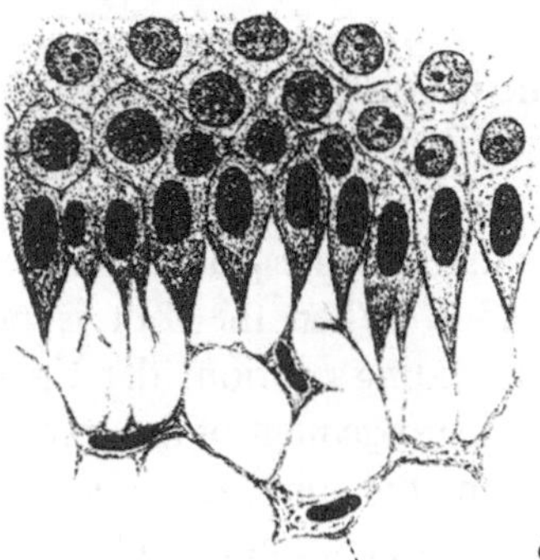
672

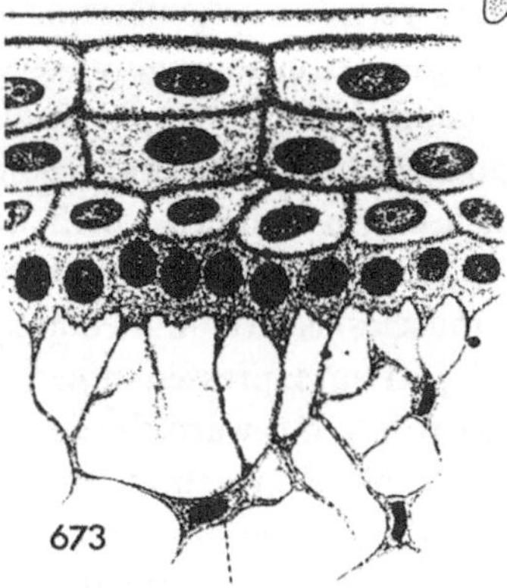
673

675

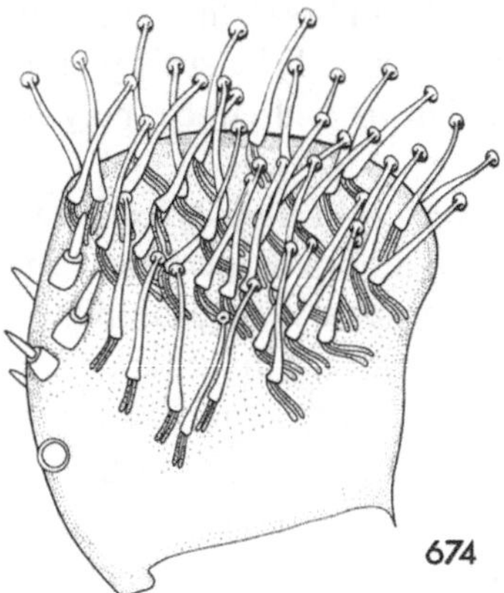
674

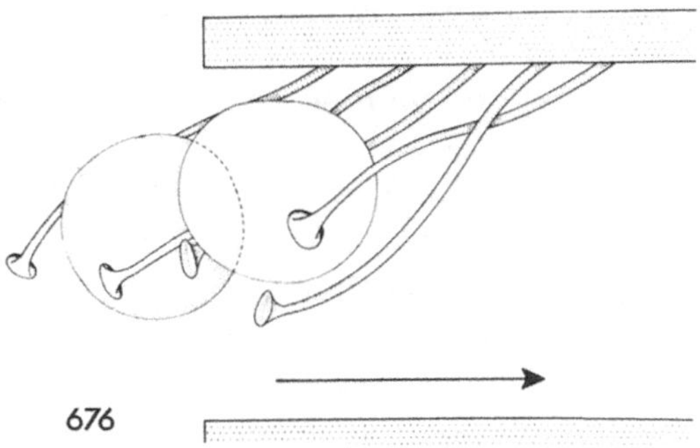
676

of which interlock with those of the epidermis. In the other amphibians this interlocking occurs by way of spiny outgrowths; but in tree frogs these projections are extended as fine threads which attach to corresponding extensions of the branched connective-tissue cells of the cuticle, forming a network (672). The epidermis of the digital pads also differs from the "typical" epidermis (673) in its thickness—the cells are cuboid rather than flattened—and in the greater degree of reticulation. The epidermal cells are further distinguished by an internal structure of numerous longitudinally directed cytoplasmic fibrils, interconnected by side branches which run obliquely or at right angles between them. In the region of the nucleus the fibrils are crowded together towards the periphery. Their parallel arrangement is especially obvious near the outer boundary of the cell.

The cuticle, apart from the reticulation mentioned above, is normally developed, with tubular glands on the plantar surface. It lies over a layer of loose connective tissue containing large lymph spaces in the terminal phalanx. Through this loose tissue run smooth-muscle fibers, sometimes in densely packed groups; the most powerful are two bundles which diverge symmetrically between the tip of the terminal phalanx and the fascia at the boundary of the plantar surface. The role of these muscles is not immediately obvious. When the foot is put in place the digital pads are pressed down with a sliding motion, the tip of the terminal phalanx rotating toward the surface; the arrangement of the smooth muscles is such that their contraction cannot account for this movement. By contracting, however, they would have the useful function of flattening the digital pad by forcing lymph out of the loose tissue, thus aiding adhesion; when they relaxed, the inflow of lymph would round out the pad and break the attachment. The longitudinal fibrils in the epidermal cells might act as little springs during the jump, and the elastic network between the epidermis and cuticle may have the same function.

The watery secretion of the glands—like any very thin fluid layer between two closely contiguous solid bodies—increases adhesion. Experiments have shown that with distilled water as a fluid layer between two glass plates, a force of about 0.7 $kdyne/mm^2$ of adhesive surface is necessary to separate them. When olive oil was used the corresponding value was 0.94 $kdyne/mm^2$, for glycerine, 1.2 $kdyne/mm^2$, and for gum-arabic solution, 1.6 $kdyne/mm^2$. With its 18 toe pads, the tree frog has available a total adhesive surface of at most 16 mm^2; since the frog weighs at most 4.5 g, even distilled water as an intermediate layer would in theory be more than adequate to ensure adhesion. The secretion of the sweat glands may well be more viscous.

Some mites live on mirror-smooth diplopods or on other animals with polished surfaces. They cannot anchor themselves there with claws, but instead fasten themselves with suckers or by direct adhesion with long (moistened?) trailing hairs (495). Trailing hairs are also found on mites which dig and live in burrows (496).

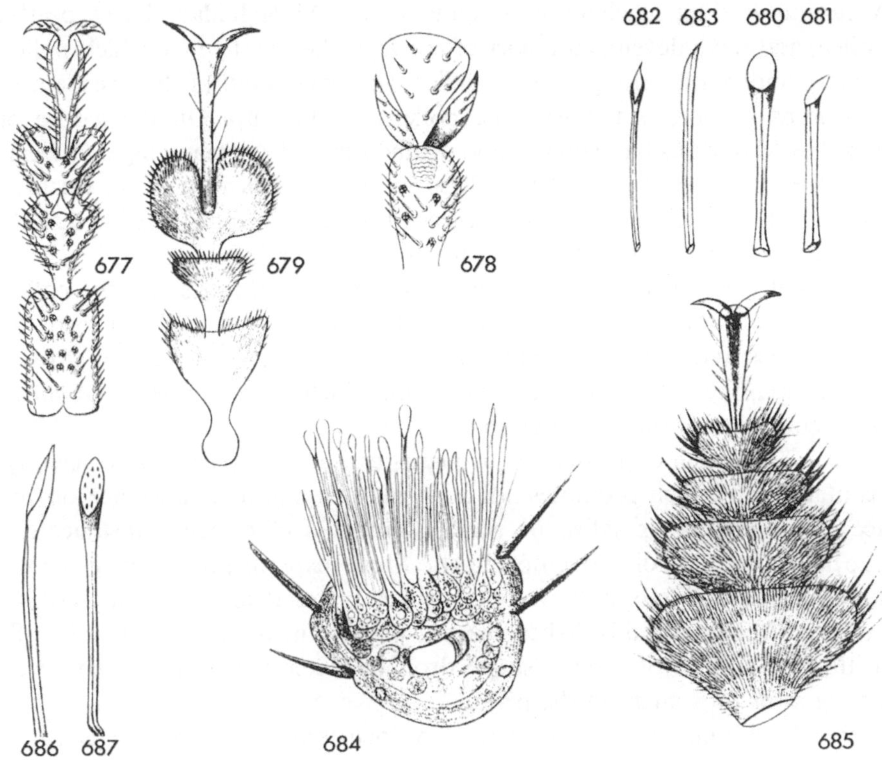

Plate 59 (Figs. 677–687)

677. Tarsus of the grasshopper Tettigonia viridissima
678. Tarsus of the grasshopper Stenobothrus flavicosta
679. Tarsus of the beetle Chrysomela tremulae
680 and 681. Chitinous tubule on the tarsus of Chrysomela tremulae
682 and 683. Chitinous tubule on the tarsus of Timarcha chloridia
684. Section through the tarsus of Hylobius abietis, showing adhesive tubules
685. Anterior tarsus of Ocypus spec.
686 and 687. Single adhesive tubule on the tarsus of Ocypus spec.

A curious feeding mechanism is found in the Malachiidae. These beetles eat pollen, and have developed a sort of conveyor-belt system by which pollen is taken up and passed along into the gullet. The only detail of this mechanism that concerns us here is the way the pollen is taken up. On the galeae of the maxillae, Malachius bipustulatus and related genera have many highly specialized bristles; because of their shape, these have been called "trumpet bristles" (674). The bristle, thick at the base, becomes thinner and more flexible distally and then expands into a funnel-shaped terminal plate (675).

When pollen is collected, the galeae are dipped several times into the pollen mass and then drawn back. Some grains of pollen are gripped between the trumpet bristles and others are held adhesively by one or more terminal discs (676). The disc, like a golf tee, has a curvature which corresponds to the mean radius of curvature of the pollen grains.

The smooth pollen of graminaceous plants can thus be picked up efficiently. The contact fluid which promotes adhesion may be a gluing substance on the surface of the pollen, or saliva or water. Remnants of cement substance are built up in the concavity of the trumpet bristle. Model experiments using ping-pong balls and correspondingly hollowed-out rubber plates demonstrate the critical importance of contact fluids. When the dry plate is inclined the ball rolls off; but with the addition of even a single drop of water it sticks securely, even during vigorous movement of the plate.

Within the chitinous sole of the tarsus of the locusts and grasshoppers (Locusta, Stenobothrus) many densely packed canals branch in a brushlike manner and finally open at the surface. There these chitinous tubules are joined in groups of three to five to form protruding pegs (677, 678); taken together they constitute an adhesive organ.

Fine chitinous tubules which secrete a fluid during walking are found among the beetles as well, particularly in the Chrysomelidae (leaf beetles), Curculionidae (weevils), and Cerambycidae (long-horn beetles). Chrysomela, for example, has these adhesive tubules only on the distal, heart-shaped tarsal segment (679), while in the genus Cassida they occur on the three proximal segments. The tubules are either markedly (Chrysomela; 680, 681) or only slightly (Timarcha; 682, 683) enlarged at the tip, which has the appearance of being cut at an angle. Single-celled, pear-shaped glands, which in Telephorus may be up to 27 μm long, discharge their secretion into these tubules. Hence there is good adhesion between substrate and the highly pliant, broadened terminal lobes of the tubules. But the rather similarly constructed tubules of the Cerambycidae and Staphylinidae (684–687), of which there are about 200,000 on the three pairs of tarsi, are said to lack an adhesive secretion; there has as yet been no experimental clarification of the adhesive mechanism of these tubules. Moreover, the function and biological significance of the fields of tubules found only on the expanded anterior tarsi of the Staphylinidae (rove beetles) has not yet been explained.

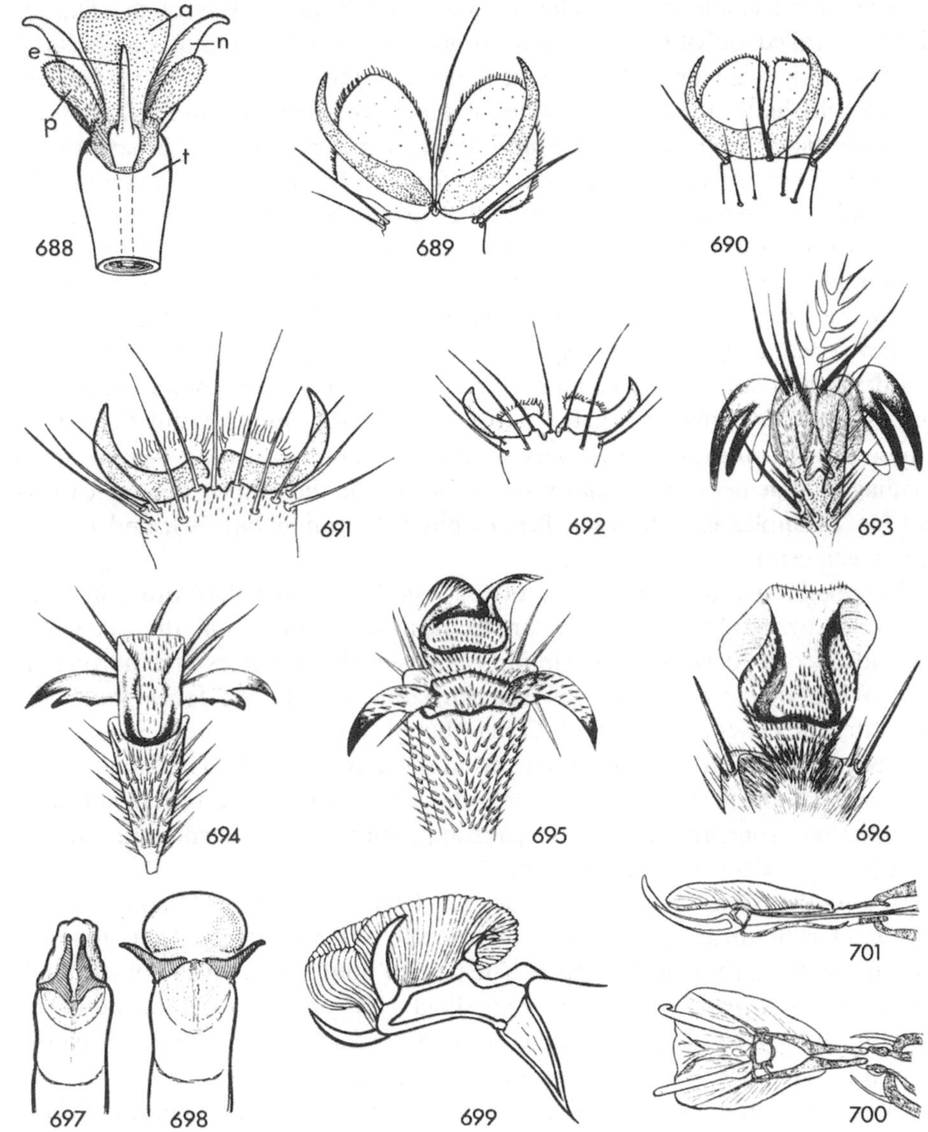

Plate 60 (Figs. 688–701)

688. Diagram of the adhesive organs on the tarsus of Diptera; e, empodium
689 and 690. Last tarsal segment of a fly living on the surface of shore deposits
691 and 692. Last tarsal segment of a fly living within shore deposits
693. End of the tarsus of Ornithomyia spec.
694. End of the tarsus of Apis mellifera
695. End of the tarsus of Vespa vulgaris
696. End of the tarsus of Vespa crabro
697 and 698. Arolium of a thysanopteran, withdrawn and everted
699. Arolium of a daddy longlegs, Scotolemon lesneri
700 and 701. Tarsus of Ixodes reduvius, from below and from the side

The characteristic lobate adhesive organs (688) on the legs of the Diptera (flies) are called pulvilli *p*; they arise from two separate chitinous platelets at the base of the claw or pretarsus *n* at the end of the tarsus *t*. In some families, for example the Stratiomyidae (soldierflies) and the Tabanidae (horseflies), there is an additional hollow lobe, the arolium *a*, in the middle of the base of the pretarsus. These adhesive lobes have the form of hairy cushions covered with a secretion. The size and significance of the various pulvilli and claws are closely related to the way of life of each kind of fly. Among flies living on the shore, for example, those species inhabiting the smooth and often dry surface of deposits left by the water are equipped with prominent pulvilli, whereas the claws are less developed (689, 690). Other species, which live deep in these deposits, where it is damp and slippery, have moderately developed pulvilli complemented by long, strong claws (691, 692); the latter are probably more useful there than adhesive or cementing organs. A very similar reduction of the lobes—and corresponding enlargement of the claws—is found in the flies which live in feathers and fur. Examples include louse flies of birds (Ornithomyia; 693) and of deer (Lipoptena cervi).

Large adhesive pads like those of the Asilidae (robber flies) are reinforced by one or two stiffening lobes of brown-pigmented chitin. On the underside, the pad is covered with small chitinous hairs which often extend beyond its edge. When the insect climbs on smooth surfaces only the lobes touch the substrate; the claws are not used. Often, though not at every step, small drops of a secretion as clear as water emerge from the adhesive pads. The hair-covered underside conforms closely to any smooth surface, and the secretion improves the adhesion. Comparable adhesive pads are found in some Hemiptera, Neuroptera (Phryganidae), and Lepidoptera.

Among the Hymenoptera, the honeybee (694), the common wasp (695), and the hornet (696) run as nimbly as flies on smooth, vertical surfaces. Their adhesive pads are particularly well developed. The garden bumblebee, on the other hand, has no adhesive pads and is correspondingly less acrobatic. Honeybees and wasps have only one large pad which evidently represents the fusion of paired primordia. This pad is especially prominent on the tarsus of Vespa (695). The edges are hairy, and adhesive secretions are produced in small quantities. These adhesive pads have a distinct central concavity. Hence accessory sucker action could be involved if the foot were first pressed down hard and then relaxed slightly, so that a reduced pressure resulted beneath the pad due to its own elastic deformation.

The Thysanoptera (thrips) can evert a large pretarsal arolium by means of blood pressure (697, 698); this serves as an adhesive "blister" in place of the two adhesive pads. Adhesive organs comparable to those of the insects also occur in other arthropods. The young of "daddy longlegs" (Phalangida, Arachnida) have an arolium in addition to the principle and accessory claws. This is a soft adhesive vesicle (699), probably distensible by blood pressure,

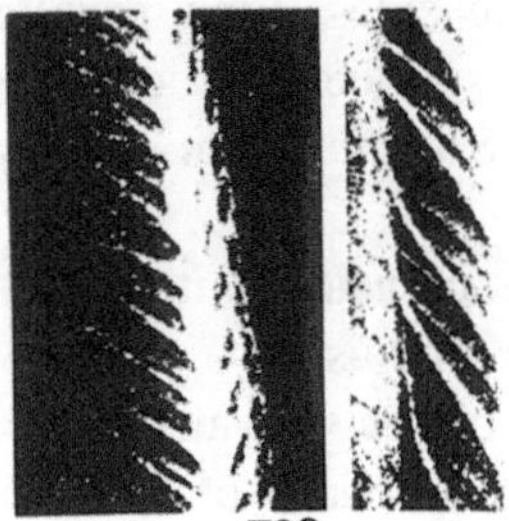
702

703

704

705

Plate 61 (Figs. 702–705)

702–704. Tarsal hairs of Cupiennius saliei, reproduced from electronmicrographs

705. Tarsal hair of Cupiennius saliei, cross section showing fringed border

with a flattened ventral surface and curved walls at the sides. Fine parallel ridges run from the base of this organ to its attachment surface. The adhesive vesicle on the tarsus of the pedipalp in the Solpugida (= Solifugae, sun spiders) is built similarly.

Extensively expanded pads are found in most of the parasitic mites. Figs. 700 and 701 show two views of the tarsal pad of Ixodes reduvius.

The tarsi of the earwig Forficula, which can walk up glass walls like a fly, are covered with extremely fine chitinous hairs. The undersides of the tarsi of the spider Cupiennius saliei also have a fine, furry coating. The individual hairs are only 2–3 μm thick; they are hollow, round at the base and flattened toward the end, with a feathery appearance (702–704) under the light microscope. According to recent electron-microscopic investigations, the feather-like projections along the hairs form characteristic sawtooth borders or fringes, but are not barbed. Taken together, the hairs form a peltlike meshwork (705) which is evidently an adhesion organ. It is not yet known whether there are fluid-secreting glands in the foot.

Among the water beetles, the males of a few dytiscid genera lack the suckers typical of this family. In their place one finds lamellae which, to judge by their shape and construction, probably have little to do with suction; they are surely organs of adhesion. The genera with these lamellar organs are Ilybius, Laccophilus, Colymbetes, and Cybister.

The lamellae of the beetle Ilybius fuliginosus, in the shape of a truncated ellipsoid (706), represent a transition between the round sucker and the extended adhesive lobes. Those of Laccophilus (707) have the typical extended form. A few of these are located on little clublike stalks at the tips of the slightly expanded tarsal segments. Colymbetes pulverosus has at its disposal a total of 112 such adhesive lobes (708–710), each with a length of about 100 μm—large in comparison to those of the other beetles.

The lamellae of Cybister roeselii have been most thoroughly investigated. On the much expanded pretarsi of the male, 130 to 170 of them are arranged in four rows (711). They are relatively large, very fragile, practically without pigment, and bear parallel striations (712). At one end they are rounded; the other end is square but with fine serrations between the ends of the striae. The remarkable flexibility of these adhesive lamellae is illustrated in Figs. 713 and 714, which were drawn from rolled lamellae. The longitudinal striation lends the appearance of a corrugated-iron sheet, and evidently provides reinforcement.

In terrestrial beetles, too, there are comparable adhesive appendages in the form of leaflets and hairs, particularly among the Carabidae (ground beetles), which are closely related to the aquatic Dytiscidae.

The tarsal segments of the genus Carabus carry marginal adhesive hairs, which appear funnel-shaped at the end and have an internal supporting helix. Carabus cancellatus has altogether 200,000 such tubules, each with a sturdy

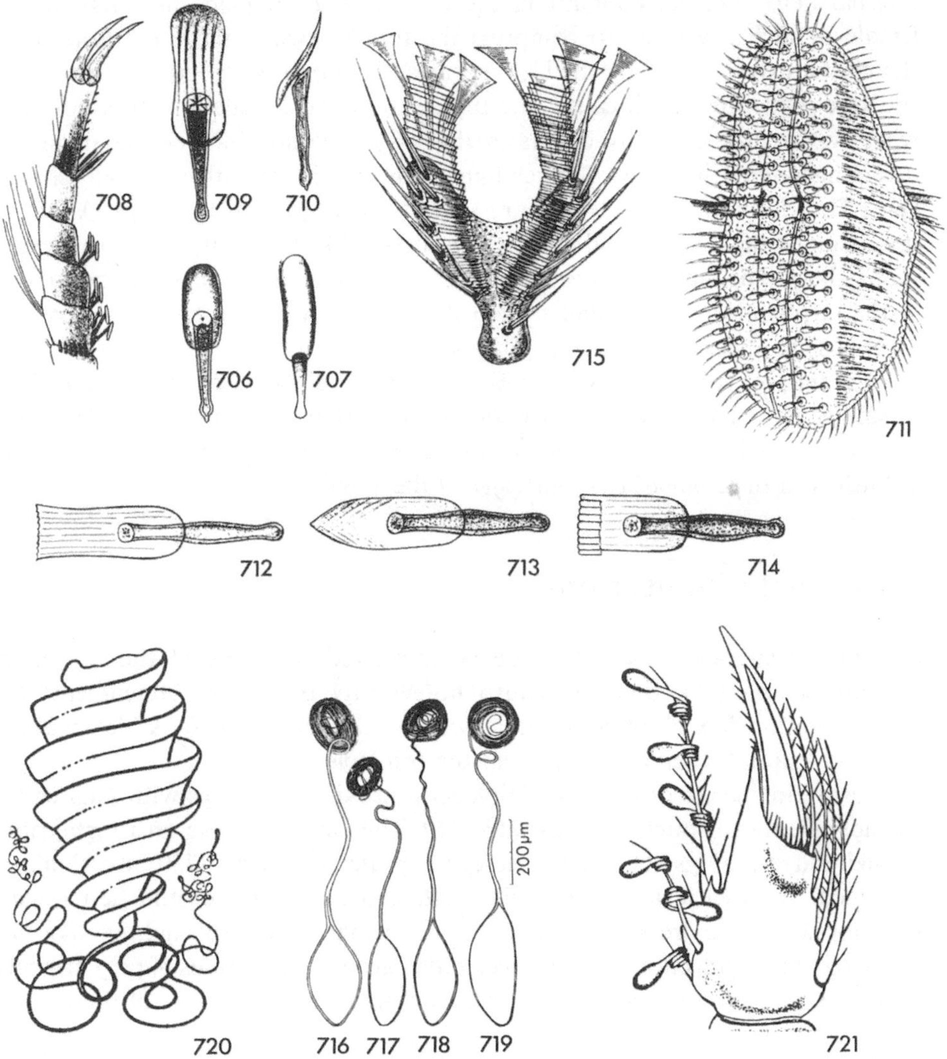

Plate 62 (Figs. 706–721)

706. Adhesive lamella of Ilybius fuliginosus
707. Adhesive lamella of Laccophilus spec.
708. Tarsus of Colymbetes pulverosus
709 and 710. Adhesive lamella on the tarsus of Colymbetes pulverosus, from below and from the side
711. Part of the tarsus of Cybister roeselii, showing adhesive lamellae
712–714. Single adhesive lamella in the natural form and rolled in various ways
715. Adhesive lamellae of Ophonus rufipes
716–719. Forms of the eggs of some European Diplozoon species, showing coiled threads: D. rutili, D. homoion, D. nagibinae, D. paradoxum
720. Egg capsule of a shark, showing coiled threads
721. Nematocysts of a fresh-water polyp, clinging to bristles of a prey animal

internal helix. During copulation, the male is said to place his tarsi on the female in such a way as to compress the tubules; when the force is lessened the elastic extension of the helical "springs" is supposed to create an internal pressure reduction which causes the tubule to attach by suction. It is not clear whether the process really occurs in this way. The species Feronia nigra has diverging longitudinal rows of much longer adhesive hairs, with elliptically shaped ends. They look like funnels and are said to have terminal openings. Although it is assumed that this animal attaches by suction like Carabus, adhesion appears more likely. The similar adhesive tubules of Nebria are located at the margin of the tarsus and those of Badister, in the center. By contrast, Ophonus rufipes (715) has spatulate adhesive appendages.

The relative roles of claws, suction, gluing and elementary adhesion in these tarsal systems have been the subject of intriguing debate for a century. Especially recommended is the lively polemic by several authors, typical of the period, published in the *Zoologischer Anzeiger* of the 1880's.

2.3 Coiled Connections

Several special cases might best be characterized in terms of the coiling of one object about another. It is also noteworthy that these attachments are among the more flexible of the examples we have chosen to classify as "detachable". Just four examples serve to document the principle.

European species of the fluke Diplozoon provide their eggs with long coiled threads like those shown in Figs. 716–719. They anchor these on irregularities in their surroundings. The elaborate egg capsules of sharks (720) are similarly equipped. Another example of coiled attachment is that of the desmoneme nematocyst of the fresh-water polyp. The nematocysts coil tightly, like the wrapped-wire connections used in electronics, around the bristles of prey animals. Escape is possible only if the prey can tear the nematocysts out of the polyp's arms, as in Fig. 721.

A fitting final example is the little-known mode of copulation of the bug Lygaeus equestris. The remarkable helical genitals of these animals and the "twisting" motions involved in their mating may even offer an entomological etymology of the curious American slang for this procedure! At the start, the male everts his penis stalk with its helical ends, and the female expands her genital aperture. The male turns the sexual member somewhat, inserts it, and then turns the body again with respect to the point of coupling. During this process, by contraction of a special bladder, the helical ends are actually twisted hydraulically into a similarly coiled recess of the female copulatory apparatus.

3. Epilogue

Of the innumerable specializations for attachment which the living world has evolved, nearly every type has been at least mentioned here, though in each case with only a few particularly compelling examples. The major omission in this guided tour, as admitted at the outset, is a survey of the joints of animals. Although it is possible in principle to classify the joints according to the notions of bioengineering, one who undertakes this task can expect to be inundated with a superabundance of structural designs. Even a moderately comprehensive summary would by itself fill another book.

From the examples of biological attachment mechanisms collected from the literature over the years, only about 25% have been selected for this volume. One can argue conservatively that the half-thousand structural types discussed here represent perhaps one percent of all the functional morphological devices which have been realized. A very wide field is open, not only to the biologist interested in mechanical design, but to the engineer as well, for the study of nature may lead him to new designs of his own.

The challenges to the bioengineer lie along a broad front; attachment mechanisms mark only the simplest end of the spectrum of complexity. An interdisciplinary approach to the functional morphology, kinematics, aerodynamics, and hydrodynamics of animal design—in general, to the physiology of movement—cannot help but be profitable to both partners in the undertaking, the biologist and the engineer. The constructions of nature are complex and sometimes difficult to understand, but in this lies their fascination.

4. Sources of the Figures

Abbreviations: HB Handbuch der Zoologie; TR Traité de Zoologie; O Original; nt not traced

1	Hyman	85	Brockhaus
2 – 3	nt	86 – 87	O
4	Hyman	88	Bronn
5 – 9	Migula	89 – 91	Heberdey
10 – 11	Kaestner	92 – 95	Lex. Technik
12 – 14	Brockhaus	96	Brockhaus
15	nt	97 – 98	Bronn
16	Wesenberg-Lund	99	Heberdey
17 – 21	Bronn	100	O
22	Kaestner	101 – 102	Peters
23 – 25	Wesenberg-Lund	103	O
26 – 31	Brockhaus	104 – 114	Hustedt
32 – 37	HB	115 – 119	Reiss
38	O	120 – 121	Meyer
39 – 44	HB	122 – 125	Jones
45 – 46	Brockhaus	126 – 134	Popp
47 – 48	Bargmann	135 – 136	O
49 – 63	Adler	137	Kaestner
64	Weber	138 – 145	Chun
65 – 66	Hustedt	146 – 150	Sterba
67	Bronn	151 – 155	Weber
68	HB	156 – 157	Meyer
69	Gruner	158 – 161	TR
70	Bronn	162	Weber
71	Gutmann	163	Bronn
72 – 73	Bronn	164 – 165	Hyman
74 – 75	Naumann	166	TR
76 – 78	Heberdey	167	Ludwig
79	Weber	168 – 169	Haeckel
80 – 81	Naumann	170 – 171	Kühn
82 – 83	Schoenichen	172	TR
84	nt	173 – 176	Haecker

177	Kosmos-Lexikon
178	Hibbs and Clark
179 – 180	TR
181 – 183	HB
184	nt
185 – 187	Handlirsch
188	O
189	TR
190	Ping
191	Keilin
192 – 193	Herkimer
194 – 197	Snodgrass
198 – 199	TR
200	Chironis
201 – 204	TR
205	Kaestner
206 – 207	Giessler
208 – 224	Tratz
225 – 238	Stanek
239 – 243	HB
244	Kaestner
245 – 246	HB
247	TR
248 – 250	HB
251 – 252	Kaestner
253	Schmidt
254 – 255	Weber
256 – 260	TR
261 – 271	Weber
272 – 280	TR
281 – 284	Palestrant
285 – 289	TR
290 – 294	Seling
295 – 296	TR
297	Palestrant
298 – 299	Ax
300 – 303	Freye
304 – 306	nt
307 – 310	Bünning
311 – 314	Brockhaus
315 – 317	HB
318 – 319	O
320 – 324	Pijl
325 – 326	HB
327	Buchsbaum and Milne
328 – 330	O
331	nt
332 – 333	Kaestner
334 – 349	HB
350	O
351 – 352	Hertel
353	Giessler
354 – 355	Hertel
356 – 359	HB
360	Kaestner
361 – 364	HB
365 – 369	TR
370 – 371	Kaestner
372	Brockhaus
373 – 374	Kaestner
375	v. Frankenberg
376	Humphries
377	Scheithauer
378	Plate
379 – 381	HB
382	Hertel
383 – 385	Herter
386 – 387	Kaestner
388 – 390	O
391 – 395	Schmidt
396 – 397	O
398 – 399	Chironis
400	O
401	Kaestner
402	Hertel
403	Chironis
404	HB
405	O
406 – 408	Schoenichen
409 – 410	HB
411 – 417	HB
418 – 419	O
420	Popp
421 – 424	O
425 – 426	HB
427	Schröder

428	Korschelt
429 – 431	Blunck
432	Törne
433	Blunck
434	TR
435	Kaestner
436	TR
437	Kaestner
438	Stümpke
439	Mohnike
440 – 443	Brosset
444 – 445	Deckart
446 – 449	Wetzel
450	nt
451 – 454	Zick
455 – 457	Beinbrech
458 – 462	Lang
463	nt
464 – 466	HB
467 – 468	Kaestner
469	Schoenichen
470 – 474	HB
475	Stahrmühlner
476 – 477	Grzimek
478	HB
479 – 481	Wesenberg-Lund
482 – 485	Rietschel
486	Hinton
487 – 493	Bischoff
494	HB
495	Kaestner
496	Kaestner
497	Schoenichen
498 – 501	Wesenberg-Lund
502	Grobben
503	Kaestner
504	Croy
505 – 511	Chun
512 – 513	Meyer
514 – 515	Brockhaus
516	Lane
517	Parker
518	Dilly et al.
519	TR
520 – 525	Bischoff
526 – 527	Bischoff
528 – 536	Nichols
537	Blunck
538 – 541	HB
542	Hyman
543 – 551	Mertens
552 – 555	Hesse-Doflein
556 – 558	HB
559 – 560	Hesse-Doflein
561	Nikolsky
562 – 563	Hesse-Doflein
564	Peters
565	Kosmos-Lexikon
566 – 568	Schoenichen
569 – 570	nt
571 – 572	Houy
573 – 574	Kosmos-Lexikon
575 – 595	Herter
596 – 598	HB
599 – 602	Schoenichen
603	Hesse-Doflein
604 – 605	Romanes and Eward
606 – 613	Mertens
614 – 615	Herter
616 – 617	Kaestner
618 – 619	HB
620	Hesse-Doflein
621	Mihail
622	HB
623 – 625	Grell
626 – 630	Lang
631	Bade
632 – 633	Nachtwey
634	TR
635	Bernard
636	Meyer
637	HB
638	nt
639 – 640	HB
641	Kaestner
642 – 644	nt

645 – 647	Kaestner
648 – 649	HB
650 – 652	TR
653 – 654	Strasburger
655	Kaestner
656	HB
657	nt
658	Bischoff
659 – 660	Kaestner
661 – 663	Wheeler
664 – 665	Hesse-Doflein
666	Schwertfeger
667	Kaestner
668 – 669	HB
670 – 671	Cochran
672 – 673	Schiberg
674 – 676	Schicha
677 – 687	Simmermacher
688	Weber
689 – 692	Remmert
693 – 696	Simmermacher
697 – 698	Weber
699	HB
700 – 701	Bronn
702 – 705	Barth
706 – 714	Simmermacher
715	HB
716 – 719	Gläser
720	Kosmos-Lexikon
721	Kaestner

References

Adler, J.: Der Mensch und sein Werkzeug. München: Südd. Verlag 1959.

Altevogt, R.: Probleme eines Fußes. Kosmos **50,** 428—430 (1954).

Alverdes: Beobachtungen an Ephemeriden- und Libellenlarven. Biol. Zentr. **43,** 577 (1924).

Ankel, W. E.: Ein rankenfüßiger Krebs mit Schaumfloß (Lepas fascicularis). Natur Volk **80,** 309—320 (1950).

Apáthy, S.: Über das Kriechen von Hirudo und Aulastoma. Zool. Jahrb. Syst. **3,** 725—294 (1888).

Apstein, C.: Bau und Funktion der Spinndrüsen der Araneida. Arch. Naturgesch. **55,** 29—74 (1889).

Assneton, R.: Loxosoma loxalina and Loxosoma saltans. Two new Species. Quart. J. Microscop. Sci. **58,** 117—143 (1912).

Audy, J. R.: Strolling on the ceiling. Malayan Nat. J. **7,** 182—190 (1953).

Ax, T.: Die Ausbildung eines Schwanzfadens in der interstitiellen Sandfauna und die Verwertbarkeit von Lebensformcharakteren für die Verwandschaftsforschung. Zool. Anz. **171,** 51—76 (1963).

Bade, E.: Das Süßwasser-Aquarium. 4. Aufl. Berlin: Pfennigsdorff 1923.

Bargmann, W., Leonhardt, G., Töndury, G.: Lehrbuch und Atlas der Anatomie des Menschen in 3 Bänden (früher: A. Rauber, F. Kopsch). 20. neubearb. Aufl., Stuttgart 1968.

Barth, R.: Die Fangdrüsen an den Beinen von Zelus (Diplodocus) leucogrammus (Perty 1834). Zool. Jahrb. Anat. **73,** 325—336 (1953).

Barth, R.: Untersuchungen an den Tarsaldrüsen von Embolyntha batesi Mac Lachlan, 1877 (Embioidea). Zool. Jahrb. Anat. **74,** 172—188 (1954).

Baudelot, E.: Note sur le disque ventral du Cyclopterus lumpus. Bull. Soc. Sci. Nat. Strasbourg **1,** 81—83 (1868).

Baumann, H.: Leime und Kontaktkleber. Berlin–Heidelberg–New York: Springer 1967.

Beck, G.: Über die Haftscheibe der Echineis remora. Inaug. Diss., Schaffhausen 1879.

Beinbrech, G.: Zur Morphologie, Fortpflanzung und Übertragung von Urceolaria paradoxa (Peritricha, Mobilina). Arch. Protistenk. **110,** 12—53 (1967).

Bernhard, C.: Über die vivipare Ephemeride Chloëon dipterum. Biol. Ctrbl. Bd. 27 (1907).

Bischoff, H.: Biologie der Hymenopteren. Eine Naturgeschichte der Hautflügler. Biologische Studienbücher V. Berlin 1927.

Bischoff, W.: Die Ökologie der paläarktischen Blepharoceridae. Ergeb. Fortschr. der Zool. **7,** 209—278 (1928).

Blunck, H.: Beitrag zu der Kenntnis der Morphologie und Physiologie der Haftscheiben von Dytiscus marginalis L. Z. Wiss. Zool. **100,** 459—492 (1912).

Bluntschli, H.: Beziehungen zwischen Form und Funktion der Primatenwirbelsäule. Gegenbaurs Morph. Jahrb. **44,** 489—517 (1912).

Bock, W. J., Wahlert, G. v.: Adaptation and the form-funktion complex. Evolution **19,** 269—299 (1965).

Böker, H.: Einführung in die vergleichende Anatomie der Wirbeltiere. 1. Bd. Jena 1935, 2. Bd. Jena 1937.

Böttcher, G.: Männliche Begattungswerkzeuge bei Sarcophaga. Deut. Entomol. Z. (1912).

Bohls, H.: Über Struktur und Wachstum der Schale von Balanus improvisus. Diss. Greifswald, 1903.

Bopp, P.: Schwanzfunktionen bei Wirbeltieren. Rev. Suisse Zool. **61,** 81—151 (1954a).

Bopp, P.: Betrachtungen zur Morphologie des Wirbeltierschwanzes. Biol. Zentr. **73,** 88—95 (1954b).

Borckert, H.: Anatomisch-physiologische Untersuchung der Haftscheibe von Cyclopterus lumpus (L.). Inaug. Diss. Kiel, 1889.

Bornemissza, G. F.: Observations on the hunting and mating behaviour of two species of Scorpion flies (Bittacidae: Mecoptera). Austr. J. Zool. **14,** 371—382 (1965).

Bowden, F. P.: Adhäsien und Reibung. Endeavour **16,** 5—18 (1957).

Brandt, H.: Untersuchungen über die Morphologie der Kopulationsorgane der Nonne Lymantria monacha L. (Lep. Lymantriidae). Zool. Anz. **114,** 161—168 (1936).

Braun, M.: Zur Bedeutung der Kutikularborsten auf den Haftlappen der Geckotiden. Arb. Zool.-Zoot. Inst. Würzburg IV (1878).

Braun, M.: Über die Haftlappen an der Unterseite der Zehen von Anolius. Arb. Zool.-Zoot. Inst. Würzburg V (1879).

Braun, M.: Über die Haftzehen der Geckos. Humboldt Monatsfr. ges. Naturwiss. 1883.

Brehm, V.: Einführung in die Limnologie. Biol. Studienbücher, Bd. X. Berlin 1950.

Bristowe, W. S.: Die Chelizeren der Spinnen, ihre Entwicklung und ihre Funktionen. Endeavour **13,** 42—49 (1954).

Brockhaus: Der Volks-Brockhaus, 10. Aufl. Leipzig 1943.

Brockhaus: Der Große Brockhaus, Wiesbaden 1953.

Brodsky, K.: Zur Kenntnis der Wirbellosenfauna der Bergströme Mittelasiens. II. Deuterophlebia mirabilis Edw. Z. Morphol. Ökol. Tiere **18,** 289—321 (1930).

Bronn, H. G.: Klassen und Ordnungen des Tierreichs. 5. Bd. Arthropoda, III. Abteilung: Insecta. Leipzig 1955.

Brosset, A.: La biologie des Chiroptères. Paris 1967.

Buchsbaum, R., Milne, L. J.: Knaurs Tierreich in Farben: Niedere Tiere. München-Zürich 1960.

Buckup, L.: Der Kopf von Myrsidea cornicis cornicis (de Geer) (Mallophaga-Amblycera). Zool. Jahrb. Anat. **77,** 241—288 (1959).

Bünning, E.: Der tropische Regenwald. Verständliche Wissenschaft, 56. Bd. Berlin-Göttingen-Heidelberg: Springer 1956.

Canna, N.: Nature and funktions of the papillary ridges of the digital skin. Anat. Record. **119,** 449—468 (1954).

Cepède, C., Willem, V.: Observations sur Trichodinopsis paradoxa Clap.-Lachm. Bull. Sci. France et Belgique. T. XLV, Ser. 7, **3** (1911).

Chakravarty, G. K., Sain, S. K., Majumdar, G.: A new nematode Spinitectus bengalensis from the fish Notopterus notopterus. Zool. Anz. **166,** 224—228 (1961).

Chatanay, J.: Sur le tarse des Dytiscides. Essai de Morphologie comparée. In: Ann. Soc. Entom. France. Bd. LXXXIX Jahrg. 1910, p. 395—466 u. tab. VIII—XV. Paris 1911.

Chironis, N. P.: Mechanisms, linkages and mechanical controls. New York etc.: McGraw Hill 1965.

Chun, K.: Die Cephalopoden. Wissenschaftliche Ergebnisse der deutschen Tiefsee-Expedition auf dem Dampfer „Valdivia" 1898—1899, Bd. 18. Jena 1910—1914.

Clara, M.: Über den Bau des Schnabels der Waldschnepfe (Scolopax rusticola L.). Z. mikrosk.-anat. Forsch. **3** (1925).

Cochran, D. M.: Knaurs Tierreich in Farben: Amphibien. München-Zürich 1961.

Coomans, A., Heiden, A. van der: Structure and formation of the feeding apparatus in Aporcelaimus and Aporcelaimellus (Nematoda: Dorylaimida). Z. Morph. Ökol. Tiere **70,** 103—118 (1971).

Crampton, G. C.: The eulabium, mentum, submentum and gular region of insects. J. entomol. Zool. **21,** 1—15 (1928).

Crome, W.: Beschreibung, Morphologie und Lebensweise der Eucta kaestneri sp. n. (Araneae, Tetragnathidae). Zool. Jahrb. Syst. **82,** 423—452 (1954).

Crome, W.: Bau und Funktion des Spinnapparates und Analhügels, Ernährungsbiologie und allgemeine Bemerkungen zur Lebensweise von Uroctea durandi (Latreille); (Araneae, Urocteidae). Zool. Jahrb. Syst. **85,** 571—606 (1957).

Dabelow, S.: Zur Biologie der Leimschleuderspinne Seytodes thoracica (Latreille). Zool. Jahrb. Syst. **86,** 85—126 (1958).

Dänzer, H.: Die Rolle des Modells und des bildhaften Denkens in der naturwissenschaftlichen Forschung. Physik-Bl. **16,** 305—309 (1960).

Dahl, F.: Beiträge zur Kenntnis des Baues und der Funktion der Insektenbeine. Arch. Naturgesch. **5** (1), 146—193 (1884).

Dahl, F.: Über den Bau und die Functionen des Insectenbeines. Zool. Anz. **7,** 38—41 (1884).

Dean, B.: A bibliography of fishes. Ch. R. Eastman (Edit.) New York Am. Mus. Nat. Hist. 1916, 1917, 1923.

Deckart, M.: Bilder vom Süßwasserpolypen Hydra. Natur Volk **88,** 236—244 (1958).

Dellit, W.-D.: Zur Anatomie und Physiologie der Geckozehe. Zena. Z. Naturw. **68,** 613—656 (1934).

Dellit, W.-D.: Zum Haftproblem der Geckoniden. Deut. Aquar. Terr. Z. **2,** 56—58 (1949).

Dewitz, H.: Die Befestigung durch einen klebenden Schleim beim Springen gegen senkrechte Flächen. Zool. Anz. **6,** 273—274 (1883).

Dewitz, H.: Über die Fortbewegung der Thiere an senkrechten glatten Flächen vermittels eines Secretes. Zool. Anz. **7,** 400—405 (1884).

Dewitz, H.: Weitere Mitteilungen über das Klettern der Insecten an glatten senkrechten Flächen. Zool. Anz. **8,** 157—159 (1885).

Dewitz, H.: Über die Fortbewegung der Thiere an senkrechten glatten Flächen vermittels eines Secretes. Pflügers Arch. ges. Physiol. **33.**

Dilly, N., Nixon, M., Packard, A.: Forces exerted by Octopus vulgaris. Pubbl. Staz. Zool. Napoli **34,** 86—97 (1964).

Dittler, R., Joos, G., Korschelt, E., Linck, G., Oltmanns, F., Schaum, K.: Handwörterbuch der Naturwissenschaften. 10 Bde. Jena 1931—1935.

Dobson, G. E.: Monograph of the Asiatic Chiroptera and catalogue of the species of bats in the collections of the Indian Museum. Calcutta 1876.

Dressler, F.: Modellbau von drei Ephemeropterenlarven. Präparator **11,** 167—172 (1965).

Dullemeijer, P.: Introduction to the symposium: principles and problems in functional anatomy. Folia Biotheor. (B) **6,** 1—4 (1966).

Eastham, L. E. S., Eassa, Y. E. E.: The feeding mechanism of the Butterfly Pieris brassicae L. Phil. Trans. B, **239** (1955).

Ehlers, E.: Die Krätzmilben der Vögel. Z. Wiss. Zool. **23** (1873).

Ehlers, E.: Beiträge zur Kenntnis der Verticalverbreitung der Borstenwürmer im Meere. Z. Wiss. Zool. **25,** 1—102 (1875).

Eichler, W.: Die Biologie der Federlinge. J. Ornithol. **84,** 471—505 (1936).

Eidmann, H.: Lehrbuch der Entomologie. Berlin: Paul Parey 1941.

Eimer, T.: Ueber fadenspinnende Schnecken. Zool. Anz. **1,** 123—124 (1878).

Emery, E.: Fortbewegung von Tieren an senkrechten und überhängenden glatten Flächen. Biol. Zentr. **4,** 438—443 (1885).

Engelmann: Nach Untersuchungen von Th. W. van Lidth de Jeude mitgetheilt von Prof. Th. W. Engelmann in Utrecht: Zur Anatomie und Physiologie der Spinndrüsen der Seidenraupe. Zool. Anz. **1,** 100—102 (1878).

Ernst, V., Ruibal, R.: The structure and development of the digital lamellae of Lizards. J. Morphol. **120,** 233—265 (1966).

Euschers, H.: Das Chitinskelett von Dytiscus marginalis. Diss. Marburg (1910).

Evans, M. E. G.: A comparative account of the feeding methods of the beetles Nebria brevicollis (F.) (Carabidae) and Philonthus decorus (Grav.) (Staphylinidae). Trans. Roy. Soc. Edinburgh **55,** 91—199 (1964).

Farooqui, M. N.: On a known and some new species of Insect nematodes. Zool. Anz. **168,** 176—300 (1962).

Feder, H. M.: On the methods used by the starfish Pisaster ochraceus in opening three types of bivalve molluscs. Ecology **36**, 764—767 (1955).

Fiebinger, J.: Über die Bauchflossen der Gobii. Anat. Anz. **27**, 140—155 (1905).

Fischer, O.: Physiologische Mechanik. Enzykl. math. Wiss. 4: Leipzig 1904.

Floericke, K.: Die Säugetiere des deutschen Waldes. Stuttgart 1908.

Floericke, K.: Kriechtiere und Lurche Deutschlands. Stuttgart 1909.

Fraenkel, G.: Biologische Beobachtungen an Janthina. Z. Morphol. Ökol. Tiere **7**, 597—608 (1927).

Fraenkel, G.: Die Grabbewegung der Soleniden. Z. vergl. Physiol. **6**, 167—220 (1927).

Franke, H.: Der Fangapparat von Chydorus sphaericus. Z. Wiss. Zool. **125**, 271—298 (1925).

Frankenberg, G. v.: Die Schere der Florfliegenpuppe (Neuropt.). Mikrokosmos **44**, 73—76 (1955).

Freeborn, S. R.: The terminal abdominal structures of male mosquitoes. Am. J. Hyg. **4**, 188—212 (1924).

Freye, H.-A.: Das Tierreich. (Redigiert von H. v. Lengerken) VII/5 Vögel. Sammlung Göschen Bd. 869, Berlin: De Gruyter 1960.

Frieling, H.: Funktionell-anatomische Untersuchungen am Scherenschnabel (Rynchops nigra intercedens Saunders). J. Ornithol. **84** (1936).

Frisch, K. v.: Hummeln als unfreiwillige Transportflieger. Natur Volk **82**, 171—174 (1952).

Fryer, G.: Studies on the functional morphology and feeding mechanism of Monodella argentarii Stella (Crustacea: Thermosbaenacea). Trans. Roy. Soc. Edinburgh **66**, 49—90 (1964).

Füller, H.: Das Thorakalskelett von Boreus westwoodi Hag. Zool. Jahrb. Anat. **73**, 423—449 (1954).

Fulton, J.: Trichodina pediculus and a closeley related species. Proc. Boston Soc. Nat. Hist. **37**, 1—29 (1923).

Gans, C.: Functional components versus mechanical units in Descriptive Morphology. J. Morphol. **128** (3), 365—368 (1969).

Geitler, L.: Die sexuelle Fortpflanzung der pennaten Diatomeen. Biol. Rev. **32**, 261—295 (1957).

Giessler, A.: Biotechnik. Leipzig: Quelle & Meyer 1939.

Gläser, H. J.: Eine neue Diplozoon-Art (Plathelminthes, Monogenoidea) von den Kiemen der Plötze, Rutilus rutilus (L.). Zool. Anz. **178**, 333—342 (1967).

Glatz, L.: Zur Biologie und Morphologie von Oecobius annulipes Lucas (Araneae, Oecobiidae). Z. Morphol. Ökol. Tiere **61**, 185—214 (1967).

Goerttler, K.: Funktion und Form. Deut. Med. Wschr. 905—914 (1968).

Golden, H. M.: Die kauenden Insektenmundteile und ihre Beziehung zu ihrer Nahrung. Inaug.-Diss. Berlin, 1926.

Goldman, E. H.: Comparisons of the mouth-parts of adult longhorn beetles with reference to their food. Trans. Amer. Entom. Soc. Philadelphia **59** (1933).

Gosse, Ph. H.: Clasping organs etc. of the Lepidoptera. Trans. Linn. Soc. London. 2. Ser. **2.** Zool. (1882).

Graber, V.: Die Insekten. München (1877).

Graber, V.: Ueber die Mechanik des Insektenkörpers. Biol. Zentr. **4**, 560—570 (1885).

Graber, V.: Die äußeren mechanischen Werkzeuge der Wirbeltiere. Leipzig und Prag 1886.

Graetz, E.: Messung der Saugkraft des Blutegels Hirudo medicinalis. Biol. Zentr. **54**, 552—553 (1934).

Grassé, P. P.: Traité de Zoologie. VI. Onychophores, Tardigrades, Arthropodes, Trilobitomorphes, Chélicérates. Paris 1949.

Grassé, P. P.: Traité de Zoologie. IX. Insectes, Paléontologie, Géonémie, Insectes Inférieurs, Coléoptères. Paris 1949.

Grassé, P. P.: Traité de Zoologie. X. 1. Fasc. Insectes supérieures et Hémipteroides. Paris 1951.

Grassé, P. P.: Traité de Zoologie. X. 2. Fasc. Insectes supérieurs et Hémipteroides. Paris 1951.

Grassé, P. P.: Traité de Zoologie. XVII. 1. Fasc. Mammifères. Les ordres: Anatomie, Ethologie, Systématique. Paris 1955.

Grassé, P. P.: Traité de Zoologie. XVII. 2. Fasc. Mammifères. Les ordres: Anatomie, Ethologie, Systématique. Paris 1955.
Greite, W.: Die Strukturbildung der Vogelfeder und ihre Pigmentierung durch Melanine. Z. Wiss. Zool. **145**, 282—336 (1934).
Grell, K. G.: Protozoologie. Berlin-Göttingen-Heidelberg: Springer 1956.
Grobben, C.: Morphologie des Fußes der Heteropoden. Arb. Zool. Inst. Wien. tom. **7** (1888).
Gruber, C.: Äußere Genitalien bei Cavia cobaya. Morph. Jahrb. **36** (1907).
Gruhl, K.: Paarungsgewohnheiten der Dipteren. Z. Wiss. Zool. **122**, 205—280 (1924).
Gruner, H. E., Deckert, K.: Das Tierreich. (Redigiert von H. v. Lengerken) IV/1 Krebse. Sammlung Göschen, Band 443, Berlin: De Gruyter 1956.
Grzimeks Tierleben. Enzyklopädie des Tierreiches. 10. Band: Säugetiere 1. Hrsg. W. Fiedler, W. Gewalt, B. Grzimek, D. Heinemann, K. Herter, E. Thenius. Zürich 1967.
Günther, K., Deckert, K.: Vergleichende Untersuchungen über Gelenkung und Muskulatur der Beine bei Orthopteren. Mitt. Zool. Mus. Berl. **19**, 377 (1933).
Gutmann, W. F.: Die Siedlungsweise der Seepocke Balanus balanus. Natur Museum **91**, 171—178 (1961).
Gutmann, W. F.: Beobachtung zum Formproblem der Seepocken-Schale. Natur Museum **92**, 193—200 (1962).
Haas, W.: Ein Fixationsmechanismus bei der Wirtsfindung der Cercarie von Diplostomum spathaceum (Trematoda). Naturwissenschaften **58**, 272—273 (1971).
Haase, A.: Untersuchungen über den Bau und die Entwicklung der Haftlappen bei den Geckotiden. Diss. Berlin, 1900.
Haeckel, E.: Report on the Radiolaria, collected by H. M. S. Challenger during the years 1873—1876. London 1887.
Haecker, V.: Tiefsee-Radiolarien (Tafeln).
Haempel, O.: Fische. Sammelband: Hilzheimer, M.: Biologie der Wirbeltiere. Ca. 1912. Ber. Bayr. Versuchsstation Fischerei München, Heft 1, 1908.
Haller, G.: Die Chitinorgane an der Saugplatte der Vorderfüße bei den Dytiscusmännchen. Arch. Naturgesch. **44**, I, 96—99 (1878).
Handbuch der Zoologie. Eine Naturgeschichte der Stämme des Tierreiches. Gegr. von Willy Kükenthal. Unter Mitarbeit zahlreicher Fachgelehrter, hrsg. von Thilo Krumbach. 4. Bd. Zweite Hälfte: Insecta. Berlin 1933—1936.
Handschin, E.: Praktische Einführung in die Morphologie der Insekten. Berlin: Borntraeger 1928.
Hannemann, H. J.: Zur funktionellen Anatomie des männlichen Kopulationsapparates von Argynnis paphia (L.). Zool. Anz. **152**, 266—274 (1954).
Hannemann, H. J.: Die Kopfmuskulatur von Micropteryx calthella (L.) (Lep.). Morphologie und Funktion. Zool. Jahrb. Anat. **75**, 177—206 (1956).
Hanstein, R. v.: Biologie der Tiere. Leipzig 1913.
Hase, A.: Beobachtungen über das Verhalten, den Herzschlag sowie den Stech- und Saugakt der Pferdelausfliege Hippobosca equina L. (Dipt. Pupipara). Z. Morphol. Ökol. Tiere **8**, 187—240 (1927).
Hassenstein, B.: Über den Funktionsbegriff des Biologen. Stud. Gen., Berlin **2**, 21—28 (1949).
Haubold, S.: Über eine neue Form sitzender Pedizellarien bei Seesternen. Zool. Anz. **103**, 199—205 (1933).
Heberdey, R. F.: Beiträge zum Bau des Subelytralraumes und zur Atmung der Coleopteren. Z. Morphol. Ökol. Tiere **33**, 667—734 (1937).
Heiberg, J.: Ueber die Drehungen der Hand. Biol. Zentr. **4**, 444—447 (1885).
Helmcke, J.-G., Otto, F.: Lebende und technische Konstruktionen. Bemerkungen zu Schalen und Raumtragwerken in Natur und Technik. Deut. Bauzeitung **67** (11), 855—861 (1962).
Hennig, W.: Der männliche Kopulationsapparat der Dipteren. III Psychodidae und Asilidae. Zool. Anz. **114**, 177—186 (1936).

Herkimer, H.: The engineer's illustrated thesaurus. New York: Chemical publishing Co. 1952.

Hertel, H.: Biologie und Technik. Mainz: Krauskopff 1963.

Herter, K.: Reizphysiologische Untersuchungen an der Karpfenlaus (Argulus foliaceus L.). Z. vergl. Physiol. **5**, 283—370 (1927).

Herter, K.: Das Tierreich (Chordatiere). VII/3 Lurche. Sammlung Göschen Bd. 847. Berlin: De Gruyter 1955.

Herter, K.: Das Tierreich. VII/4 Kriechtiere. Sammlung Göschen 447/447 a, Berlin 1960.

Hesse, R., Doflein, F.: Tierbau und Tierleben in ihrem Zusammenhang betrachtet. 2. Aufl. 1935—1943. 1. Bd.: Der Tierkörper als selbständiger Organismus. Jena 1935. 2. Bd.: Das Tier als Glied des Naturganzen. Jena 1943.

Hiller, U.: Untersuchungen zum Feinbau und zur Funktion der Haftborsten von Reptilien. Z. Morphol. Ökol. Tiere **62**, 307—362 (1968).

Hiller, U., Blaschke, R.: Zum Haftproblem der Geckofüße. Naturwissenschaften **54**, 344—345 (1967).

Hoerschelmann, H.: Schnabelform und Nahrungserwerb bei Schnepfenvögeln (Charadriidae und Scolopacidae). Zool. Anz. **184**, 302—327 (1970).

Hofeneder, H.: Über die Larven der Blepharoceriden und ihren merkwürdigen Anheftungsapparat. Verhandl. Zool.-Botan. Ges. Wien **77**, 82—98 (1927).

Hofer, H.: Über das Bewegungsspiel der Klammeraffen. Natur Volk **88**, 397—412 (1958).

Honio, I.: Zum Mechanismus der Gehbewegung beim Blutegel. Annotat. Zool. Japan 1936.

Hora, S. L.: The adhesive apparatus on the toes of certain Geckos and tree-frogs. J. Proc. Asiatic Soc. Bengal, **9**, 137—145.

Hora, S. L.: V. Ecology, Bionomics and Evolution of the torrential Fauna with special reference to the organs of attachment. Phil. Trans. B. **218**, 172—282 (1930).

Hornberger, F.: Die Copula der Aeschna cyanea. Jen. Z. Nat. Wiss. **55** (1918).

Houy, R.: Beiträge zur Kenntnis der Haftscheibe bei Echeneis. Zool. Jahrb. Anat. **29**, 101—138 (1910).

Howard, T. E.: Rapid excavation. Sci. Am. **217**, 74—84 (1967).

Humphries, D. A.: Function of combs in Ectoparasites. Nature **215** (Nr. 5098) 319 (1967).

Hustedt, F.: 10. Bacillariophyta (Diatomeae). 2. Aufl. Bearbeitet v. Friedrich Hustedt. 1930. In: Die Süsswasser-Flora Mitteleuropas. Hrsg. Adolf Pascher. Jena: Fischer 1927.

Hustedt, F.: Die Süßwasser-Flora Mitteleuropas. Heft 10: Bacillariophyta (Diatomae). 2. Aufl. Jena: Fischer 1930.

Hyman, L. H.: The Invertebrates: Platyhelminthes and Rhynchocoela. The acoelomate Bilateria. Vol. II. New York-Toronto-London 1951.

Hyman, L. H.: The Invertebrates: Echinodermata. The coelomate Bilateria. Vol. IV. New York-Toronto-London 1955.

Ilg, L.: Über larvale Haftorgane bei Teleosteern. Zool. Jahrb. Anat. **72**, 577—600 (1952).

Illies, J.: Die Lebensgemeinschaft des Bergbaches. Die Neue Brehm-Bücherei 289. Wittenberg Lutherstadt 1961.

James-Clark, H.: Tha anatomy and physiology of the Vorticellidan parasite (Trichodina pediculus Ehr.) of Hydra. Mem. Bonston Soc. Nat. Hist. **1**, 114—130 (1866).

Jenkin, P. M.: The filter feeding and food of flamingo. Phil. Trans. **240**, 401—493 (1957).

Jensen, D.: The hagfish. Sci. Am. **214**, 82—90 (1966).

Jones, J. C., Wheeler, R. E.: An analytical study of Coitus in Aedes aegypti (Linnaeus). J. Morph. **117**, 401—423 (1965).

Jones, J. C.: The sexual life of a mosquito. Sci. Am. **218**, 108—116 (1968).

Joseph, G.: Vorläufige Mittheilungen über Innervation und Entwicklung der Spinnorgane der Insecten. Zool. Anz. **3**, 326—328 (1880).

Kämpfe, L., Kittel, R., Klapperstück, J.: Leitfaden der Anatomie der Wirbeltiere. Berlin 1955.

Kaestner, A.: Lehrbuch der speziellen Zoologie. Teil I: Wirbellose, 1. Halbb. Stuttgart 1954/55.

Kaestner, A.: Lehrbuch der speziellen Zoologie. Teil I: Wirbellose, 2. Halbb. Stuttgart 1963

Kaltenbach, A.: Gifte und Giftwaffen im Tierreich. Naturw. Rundschau **24**, 380—388 (1971).

Karg, W.: Systematische Untersuchung der Eviphididae Berlese, 1913 (Acarina, Parasitiformes) mit einer neuen Art aus Ackerböden. Zool. Anz. **170**, 269—281 (1963).

Kaszab, Z.: Merkmale der Adaptation, Spezialisation, Konvergenz, Korrelation und Progression bei den Meloiden (Col.). Acta Zool. Acad. Sci. hung (Budapest) **9**, 135—175 (1963).

Kemper, H.: Beiträge zur Biologie der Bettwanze (Cimex lectularius L.) III. Über den Mechanismus des Stech-Saugaktes. Z. Morphol. Ökol. Tiere **24**, 491—518 (1932).

Kepner, W. A., Reynolds, B. D., Goldstein, L., Taylor. J. H.: The structure, development and discharge of the penetrant of Pelmatohydra oligactis (Pall.) J. Morphol. **72**, 561—587 (1943).

Kepner, W. A., Reynolds, B. D., Godstein, L., Britt, G.: The discharge of nematocystes of Hydra with special reference to the penetrant. J. Morphol. **88**, 23—48 (1951).

Kerkut, G. A.: The forces exerted by the tube feet of the starfish during locomotion. J. Exp. Biol. **30**, 575—583 (1953).

Kerkut, G. A.: The mechanisms of coordination of the starfish tube feet. Behaviour **6**, 206—233 (1954).

Kerkut, G. A.: The Retraction and Protraction of the Tube Feet of the Star fish. Behaviour **8**, 112—130 (1955).

Kirchner, O.: Blumen und Insekten. Leipzig: J. B. Teubner 1911.

Knoll, F.: Insekten und Blumen. Experimentelle Arbeiten zur Vertiefung unserer Kenntnisse über die Wechselbeziehungen zwischen Pflanzen und Tieren. Abhandl. Zool.-Bot. Ges. Wien **12**, 1—646 (1921—1926).

Komarek, J.: Die Morphologie und Physiologie der Haftscheiben der Blepharoceridenlarven. S.-B. Böhm. Ges. Wiss. **25**, 1—28 (1914).

Korschelt, E.: Der Gelbrand Dytiscus marginalis. Leipzig: Engelmann 1923—1924.

Kosmos-Lexikon der Naturwissenschaften. 2 Bde. Stuttgart 1953.

Kramer, P.: Über eine neue Pelzmilbe des Bibers (Haptosoma truncatum nov. gen. nov. sp.). Zool. Anz. **19**, 134—136 (1896).

Krefft, G.: Ichthyologische Mitteilungen aus dem Institut für Seefischerei der Bundesanstalt für Fischerei. I. Zool. Anz. **150**, 275—282 (1953).

Krieg, H.: Über die mechanisch wirkenden Papillen der Kaninchenzunge. Z. Anat. Entwicklungsgeschichte 61 (1921).

Krompecher, S.: Form und Funktion in der Biologie. Kl. naturw. Bibl. Reihe Biol. 2. Leipzig: Akad. Verlagsges. Gees & Portig 1966.

Krumbiegel, I.: Biologie der Säugetiere. 1. Bd. Krefeld 1953. 2. Bd. Krefeld-Baden-Baden 1955.

Kühn, A.: Morphologie der Tiere in Bildern. Berlin 1921.

Kugler, H.: Einführung in die Blütenökologie. Stuttgart: G. Fischer 1955.

Kuhl, W.: Untersuchungen über die Bewegungsphysiologie der Fangorgane am Kopf der Chaetognathen. Z. Morphol. Ökol. Tiere **24**, 526—575 (1932).

Kullmann, E.: Spinnorgan mit 40000 „Düsen". Umschau **69**, 82—83 (1969).

Kunitzky, I.: Die Zeit der Entstehung der Borsten und Mechanismus der Bewegung bei den Geckotiden. Bull. Acad. Imp. Sci. St.-Petersbourg **18**, 21—24 (1903).

Lacombe, D., Liquori, V. R.: Comparative histological studies of the cement apparatus of Lepas anatifera and Balanus tintinnabulum. Biol. Bull. **137**, 170—180 (1969).

Lane, F. W.: Kingdom of the octopus. London: Jarrolds 1957.

Lang, A.: Handbuch der Morphologie der wirbellosen Tiere. Jena 1912—1913.

Langer, K.: Über den Gelenkbau bei den Arthrozoen. Denkschr. d. Akad. d. Wiss. zu Wien, Physikmath. Kl. 18 (1860).

Lautenschlager, F.: Über eine besondere Anpassung mechanischer Zungenpapillen an die Nahrungsaufnahme im Fluge bei Fledermäusen. Zool. Anz. **109**, 263—266 (1935).

Lavoie, M. E.: How sea stars open bivalves. Biol. Bull. **111**, 114—122 (1956).

Levy, F.: Copula von Spiola atlantica. Zool. Anz. **39** (1912).

Lieberkind, J.: Über die Haftorgane bei Jungen von Pterophyllum eimekei. E. Dahl. Zool. Anz. **97**, 55—61 (1932).

Lindner, E.: Beiträge zur Kenntnis der Larven der Limoniidae (Diptera). Z. Morphol. Ökol. Tiere **48**, 209—319 (1959).

Linstow, V.: Neue Art der Copula bei Distomen. Zool. Anz. **28** (1905).

List, Th.: Morphologisch-biologische Studien über den Bewegungsapparat der Arthropoden. I. Astacus fluv. Morph. Jb. **22**.

Lowne, B.: On the so-called Suckers of Dytiscus and the Pulvilli of Insects. In: Monthly Microscopical Journal: Trans. Roy. Micr. Soc. London. **V**, 267—271, Tab. LXXXIX.

Ludwig, H.: Seesterne. Anvers: Buschmann 1903.

Ludwig, W.: Symmetrieforschung im Tierreich. Stud. gen., Berlin **2**, 231—239 (1949).

Maderson, P. F. A.: Keratinized epidermal derivatives as an aid to climbing in gekkonid lizards. Nature (London) **203**, 780—781 (1964).

Maderson, P. F. A.: How does the gecko get a grip? New Scientist **406**, 509 (1964).

Märkel, K., Meyer, I.: Zur Systematik der deutschen Euphtiracarini (Acari, Oribatei). Zool. Anz. **163**, 327—342 (1959).

Mahendra, B. C.: Contributions to the bionomics, anatomy, Reproduction and development of the Indian house-gecko, Hemidactylus flaviviridis Rüppel. Part II, The problem of locomotion. Proc. Ind. Acad. Sci. **4**, 288—306 (1941).

Mannheims, B. J.: Beiträge zur Biologie und Morphologie der Blepharoceriden (Dipt.). Zool. Forsch. **2**, 1—114 (1935).

Manton, S. M.: Habits of life and evolution of body design in Arthropoda. J. Linn. Soc. London **44**, 58—72 (1958).

Manton, S. M. (Mrs. J. P. Harding): Mandibular mechanisms and the evolution of Arthropods. Phil. Trans. **247**, 1—184 (1964).

Marcus, E.: Zur vergleichenden Anatomie und Histologie der Tardigraden. Zool. Jahrb. Physiol. **45**, 99—158 (1928).

Marcus, H.: Über die Atmungsorgane bei Tracheaten. I. Allgemeines über die Atmung der Insekten. Z. Wiss. Zool. **160**, 165—185 (1956).

Marinelli, W.: Über den Schädel der Schnepfe. Versuch einer funktionellen Analyse des Kieferapparates. Palaeobiologica, 1. Bd. 1. Teil, 135—160, 1928.

Marshall, A. J.: Biology and comparative physiology of birds. New York-London 1960, 1961.

Martens, E. v.: Zur Kenntnis der fadenspinnenden Schnecken. Zool. Anz. **1**, 249—251 (1878).

Martini, E.: Ein Kalkflagellat in den Meeren der Gegenwart und Vorzeit. Natur Volk **91**, 335—339 (1961).

Matthes, D.: Excitatoren und Paarungsverhalten mitteleuropäischer Malachiiden (Coleopt., Malacodermata). Z. Morphol. Ökol. Tiere **51**, 375—546 (1962).

Matthes, D., Schicha, E.: Funktionsmorphologische Studien an den Mundwerkzeugen pollenfressender Coleopteren (Malachiidae). Naturwissenschaften **53**, 364 (1966).

Meisenheimer, J.: Geschlecht und Geschlechter im Tierreiche. I. Die natürlichen Beziehungen. Jena 1921. II. Die allgemeinen Probleme. Jena 1930.

Meixner, A.: Genitalapparat der Chloroclystis rectangulata. Z. Wiss. Insekt. Biol. **2** (1906).

Mell, C.: Das Leben der Vampyrella variabilis. Mikrokosmos 1919, 1920.

Mertens, R.: Die Larven der Amphibien und ihre evolutive Bedeutung. Zool. Anz. **164**, 337—358 (1960).

Mertens, R.: Der Eidechsenschwanz als Haftorgan. Senckenbergiana Biol. **45**, 117—122 (1964).

Meyer, W. Th.: Tintenfische mit besonderer Berücksichtigung von Sepia und Octopus. Leipzig 1913.

Mickoleit, G.: Über den Ovipositor der Neuropteroidea und Coleoptera und seine phylogenetische Bedeutung (Insecta, Holometabola). Z. Morph. Ökol. Tiere **74**, 37—64 (1973).

Migula, W.: Kryptogamen-Flora, Bd. II: Allgem. Berlin: H. Behrmüller 1907.

Mihail, N.: Funktionelle Untersuchungen an den Verdauungsorganen von Euodontomyzon danfordi. Zool. Anz. **168,** 130—143 (1962).

Mikulski, J.: Beitrag zur Kenntnis des Baues und der Funktion der Haftballen des Laubfrosches (Hyla arborea L.). Extr. Bull. l'Acad. pol. Sci. et d. Lettr. (1928).

Miller, W. A.: Periodontal attachment apparatus in the young Caiman sclerops. Arch. oral Biol. **13,** 735—743 (1968).

Mitchell, R.: The structure and evolution of water mite mouthparts. J. Morph. **100,** 41—59 (1962).

Mohnike, O.: Über das Vermögen verschiedener Säugetiere sich mittels des atmosphärischen Drucks an glatten, mehr oder weniger senkrechten Flächen festhalten und aufwärtsbewegen zu können. Z. Wiss. Zool. **32,** 388 (1879).

Mollenhauer, D.: Betrachtungen über Bau und Leistung der Organismen I. Aufs. u. Reden Senckenb. Naturf. Ges. **19** (1970).

Moodie, R. L.: Clasping organs of extinct and recent Amphibia. Biol. Bull. **14** (1908).

Mound, L. A.: The feeding apparatus of thrips. Bull. ent. Res. **60,** 547—548 (1971).

Mühlenberg, M.: Die Abwandlung des Eilegeapparates der Bombyliidae (Diptera). Eine funktionsmorphologische Studie. Z. Morph. Ökol. Tiere **70,** 1—72 (1971).

Müller, H. J.: Über Bau und Funktion des Legeapparates der Zikaden (Homoptera cicadina). (Zugleich erster Teil der Beiträge zur Biologie mitteleuropäischer Zikaden.) Z. Morphol. Ökol. Tiere **38,** 534—629 (1941).

Müller, R. J.: Die Mechanik der syndaktylen Zehen von Macropus und anderen Beuteltieren und ihre Verwendung als Putzorgan. Z. Morph. Ökol. Tiere **17** (1932).

Nachtigall, W.: Einige Beobachtungen über die Fortbewegung der Dytisciden außerhalb des Wassers. Zool. Anz. **166,** 105—108 (1961).

Nachtigall, W.: Biotechnik. Heidelberg: Quelle & Meyer 1971.

Nachtwey, R.: Wunderbare Welt im Wassertropfen. Wiesbaden: Brockhaus 1913.

Naumann, H.: Der Gelbrandkäfer. Die Neue Brehm-Bücherei 162. 1. Aufl. Wittenberg Lutherstadt 1955.

Neumann, A. W.: Die Oberflächenspannung fester Körper. Umschau **6,** 198 (1967).

Newman, H. H.: Contact organs in the Killifishes. Biol. Bull. **17** (1909).

Nichols, D.: The histology of the Tube-feet and Clavulae of Echinocardium cordatum. Quart. J. Micr. Sci. **100,** 73—87 (1959).

Nikolski, G. W.: Spezielle Fischkunde. Berlin 1957.

Nußbaum, R.: Der Thorax von Basilia nana (Diptera Nycteribiidae). Zool. Jahrb. Anat. **78,** 313—368 (1960).

Östergen, H.: Über die Function der ankerförmigen Kalkkörper der Seewalzen. Zool. Anz. **20,** 148—159 (1897).

Osterwald, T.: Beobachtungen über die Nahrungsaufnahme des Maikäfers und eine Untersuchung seiner Mundwerkzeuge. Mikrokosmos **43,** 186—188 (1953).

Otto, F.: Zugbeanspruchte Konstruktionen 1. Frankfurt/M.-Berlin: Ullstein Fachverlag 1962.

Paine, V. L.: Adhesion of the Tube Feet in Starfishes. J. Exp. Zool. **45,** 361—366 (1926).

Paine, V.: Tube feet as autonomous organs. Am. Nat. **62,** 517—529 (1929).

Palestraut, S. S.: Practical pictorial guide to mechanisms and machines. New York: University Books 1950.

Parker, G. H.: The power of suction in the sea-anemone Cribrina. J. Exp. Zool. **24,** 219—222 (1917).

Parker, G. H.: The power of adhesion in the suckers of Octopus bimaculatus verrill. J. Exp. Zool. **33,** 391—394 (1921).

Pauly, M.: Die Mundwerkzeuge der Carabiodea. Inaug.-Diss. Berlin 1915.

Pérez, Ch.: Les Rhizocéphales parasites des Pagures. Verhandl. Schweiz. Naturf. Ges. **112,** 261—276 (1931).

Peter, K.: Die Methoden der Rekonstruktion. Jena: G. Fischer 1906.

Peters, H. M.: Über larvale Haftorgane bei Silapia (Cichlidae, Teleostei) und ihre Rückbildung in der Evolution. Zool. Jahrb. Physiol. **71**, 287—300 (1965).

Peters, N. (Jr.): Opercular- und Postopercularorgan (Occipitalorgan) der Gattung Kneria (Kneriidae, Pisces) und ein Vergleich mit verwandten Strukturen. Z. Morphol. Ökol. Tiere **59**, 381—435 (1967).

Peters, R.: Die Radula einheimischer Schnecken. Mikrokosmos **60**, 261—269 (1971).

Pfau, H. K.: Struktur und Funktion des sekundären Kopulationsapparates der Odonaten (Insecta, Palaeoptera), ihre Wandlung in der Stammesgeschichte und Bedeutung für die adaptive Entfaltung der Ordnung. Z. Morph. Ökol. Tiere **70**, 281—371 (1971).

Piersig, R.: Deutschlands Hydrarachniden. Zoologica **22** (1897—1900).

Pijl, L. van der: Principles of dispersal in higher plants. 2. Aufl. Berlin-Heidelberg-New York: Springer 1972.

Pockrandt, D.: Beiträge zur Histologie der Schlangenhaut. Zool. Anz. **62**, 275—322 (1936).

Pohl, F.: Beziehungen zwischen Pollenbeschaffenheit, Bestäubungsart und Fruchtknotenbau. Untersuchungen zur Morphologie und Biologie des Pollens. I. Botan. Zentr. Beih. **46**, 247—285 (1929a).

Pohl, F.: Kittstoffreste auf der Pollenoberfläche windblütiger Pflanzen. Untersuchungen zur Morphologie zu Biologie des Pollens. I. Botan. Zentr. Beih. **46**, 286—305 (1929b).

Poletajew, N.: Über die Spinndrüsen der Blattwespen. Zool. Anz. **8**, 22—23 (1885).

Poljanec, L.: Morphologie der äußeren Geschlechtsorgane bei den männlichen Lepidopteren. Arb. Zool. Inst. Wien, tome **13** (1900).

Popham, E. J.: The feeding habits of Arixenia and Hemimerus (Dermapt.). Verhandl. 11. internat. Kongr. Entom. Wien I. p. 323—327 (1960).

Popp, E.: Die Begattung bei den Vogelmilben Pterodectes Robin (Analgesoidea, Acari). Z. Morph. Ökol. Tiere **59**, 1—32 (1967).

Porsch, O.: Windpollen und Blumeninsekten. Österr. Botan. Z. **103**, 1—18 (1956).

Postner, M.: Kopulationsverhalten bei den Gattungen Bombus und Psithyrus (Apidae, Hymenoptera). Zool. Anz. **151**, 300—306 (1953).

Raap, F.: Über die Anfertigung eines Gummi-Kork-Modells von Drosophila melanigaster. Präparator **6**, 74—77 (1960).

Rauther, M.: Der Saugmund von Discognathus. Zool. Jahrb. Physiol. **45**, 45—76 (1928).

Rees, D. M.: Morphology of the terminalia and internal reproductive organs and copulation in the mosquito, Culiseta melanura (Williston). Proc. Ent. Soc. Wash. **53**, 233—246 (1951).

Reh, L.: Über Asymmetrie und Symmetrie im Tierreich. Biol. Zentr. **19**, 625—652 (1899).

Reichardt, H.: Untersuchungen über den Genitalapparat der Asiliden. Z. Wiss. Zool. **135**, 257—301 (1929).

Reicher, E.: Über das Haftorgan von Erobius fluviatilis. Mitt. Naturf. Ges. Bern 322—347, 1912 (1913).

Reiss, F.: Zum Kopulationsmechanismus bei Chironomiden (Diptera). Chironomidenstudien IV. Zool. Anz. **176**, 440—449 (1966).

Remane, A.: Die Beziehungen zwischen Phylogenie und Ontogenie. Zool. Anz. **164**, 306—337 (1960).

Remmert, H.: Lebensformtypen von Strandfliegen (Diptera). Zool. Anz. **165**, 432—438 (1960).

Reti, L.: Leonardo on bearings and gears. Sci. Am. **224**, 100—110 (1971).

Rietschel, P.: Bau und Funktion des Wehrstachels der staatenbildenden Bienen und Wespen. Z. Morphol. Ökol. Tiere **33**, 313—357 (1937).

Rietschel, P.: Bau, Funktion und Entwicklung der Haftorgane der Blepharoceridenlarven. Z. Morphol. Ökol. Tiere **50**, 239—265 (1961/62).

Rietschel, P.: Blepharoceridenlarven. Natur Museum **94**, 188—192 (1964).

Rimsky-Korsakow, M.: Beitrag zur Kenntnis der Embiiden. Zool. Anz. **29**, 433—442 (1905).

Rimsky-Korsakow, M.: Über das Spinnen der Embiiden. Zool. Anz. **36**, 153—156 (1910).

Risler, H.: Der Kopf von Bovicolar caprae (Gurlt) (Mallophaga). Zool. Jahrb. Anat. **71**, 325—374 (1951).

Ritter-Zahony, R. v.: Zur Anatomie des Chaetognathenkopfes. Zool. Erg. Nr. XVI, Ber. Komm. Erforschung öst. Mittelmerr.) Denkschr. Akad. Wiss. Wien, Math.-naturwiss. Kl. 84 (1909).

Romanes, G. J., Ewart, J. C.: Observations on the locomotor system of Echinodermata. Phil. Trans. Roy. Soc., Part III, 829—885 (1881).

Rombouts, J. E.: Über die Fortbewegung der Fliegen an glatten Flächen. Zool. Anz. **7**, 619—623 (1884).

Ruibal, R., Ernst, V.: The structure of the digital setae of lizards. J. Morphol. **117**, 271—293 (1965).

Ruser, M.: Beiträge zur Kenntnis des Chitins und der Muskulatur der Zecken (Ixodidae). Z. Morphol. Ökol. Tiere **27**, 199—261 (1933).

Russel, E. S.: Form and Function.—A contribution to the history of animal morphology. London: John Murray 1916.

Rutschke, E.: Morphologisch-funktionelle Untersuchungen über Reibungsstrukturen an Schwanzfedern. Z. Wiss. Zool. **181**, 331—352 (1970).

Salzer, R.: Konstruktionsanatomische Untersuchungen des männlichen Postabdomens von Calliphora erythrocephala Meigen (Insecta, Diptera). Z. Morphol. Ökol. Tiere **63**, 155—238 (1968).

Sanderson, I. T.: Knaurs Tierbuch in Farben: Säugetiere. München-Zürich 1956.

Saroyan, J. R., Lindner, E., Dooley, C. A.: Repair and reattachment in the Balanidae as related to their cementing mechanism. Biol. Bull. **139**, 333—350 (1970).

Sattler, W.: Beiträge zur Kenntnis von Lebensweise und Körperbau der Larve und Puppe von Hydropsyche Pict. (Trichoptera) mit besonderer Berücksichtigung des Netzbaues. Z. Morphol. Ökol. Tiere **47**, 115—192 (1958).

Saxena, S. C., Chandy, M.: Adhesive apparatus in certain Indian hill stream fishes. J. Zool. **148**, 315—340 (1966).

Schäfer, W.: Sandkorallen. Natur Volk **79**, 244—245 (1949).

Schäfer, W.: Form und Funktion der Brachyuren-Schere. Abhandl. Senckenb. Naturf. Ges. **498** (1954).

Schaffer, J.: Anatomisch-histologische Untersuchungen über den Bau der Zehen bei Fledermäusen und einigen kletternden Säugetieren. Ein weiterer Beitrag zur Kenntnis der Bindesubstanzen. Z. Wiss. Zool. **83**, 231—284 (1905).

Scheithauer, W.: Kolibris—Fliegende Edelsteine. München: BLV o. J.

Scherney, F.: Unsere Laufkäfer. Die Neue Brehm-Bücherei H. 245. Wittenberg Lutherstadt 1959.

Schicha, E.: Eine Kehrmaschine für Blütenstaub. Mikrokosmos **55**, 293—296 (1966).

Schicha, E.: Morphologie und Funktion der Malachiidenmundwerkzeuge unter besonderer Berücksichtigung von Malachius bipustulatus L. (Coleopt., Malacodermata.). Z. Morphol. Ökol. Tiere **60**, 376—433 (1967).

Schiemenz, P.: How do starfish open oysters? J. Mar. Biol. Assoc. U. K. **4**, 266—285 (1896).

Schiemenz, H.: Vergleichende funktionell-anatomische Untersuchungen der Kopfmuskulatur von Theobaldia und Eristalis (Dipt. Culicid. und Syrphid.). Deut. Entom. Z., N. F. **4**, 268—331 (1957).

Schindler, O.: Zur Biologie der Larven von Barsch (Perca fluviatilus L.) und Hecht (Esox lucius L.). Verhandl. deutsch. Z. Ges. **37**, 141—149 (1935).

Schliemann, H.: Bau und Funktion der Haftorgane von Thyroptera und Myzopoda (Vespertilionoidea, Microchiroptera, Mammalia). Z. Wiss. Zool. **181**, 353—400 (1970).

Schmidt, E.: Vergleichende Morphologie des 2. und 3. Abdominalsegments bei männlichen Libellen. Zool. Jahrb. Anat. **39**, 87—200 (1916).

Schmidt, H.: Zur Anatomie und Physiologie der Geckopfote. Jena. Z. Naturw. **39**, 551 (1904).

Schmidt, W. J.: Die Haut der Geckoniden. Z. Wiss. Zool. 101 (1912).

Schmitt, H.: Biologische und morphologische Untersuchungen an Malachiiden (Col. Malacodermata). Entom.-Bl. (Krefeld) **41—44**, 167—177 (1949).

Schmitt, J. B.: The feeding mechanism of adult lepidoptera. Smith. Misc. Coll. **97,** 1—20 (1938) (1938).

Schoenichen, W.: Biologie und Physik. Leipzig 1909.

Schoenichen, W.: Praktikum der Insektenkunde. Jena 1918.

Schremmer, F.: Morphologische Anpassungen von Tieren – insbesondere Insekten – an die Gewinnung von Blumennahrung. Verhandl. Deut. Zool. Ges. Zool. Anz. Suppl. **25,** 375—401.

Schremmer, F.: Die Mundteile der Brachycerenlarven und der Kopfbau der Larve von Stratiomys chamaeleon L. Österr. Zool. Z. **3,** 326—297 (1951).

Schremmer, F.: Der bisher unbekannte Pollensammelapparat der Honigwespe Celonites abbreviatus VILL (Vespidae Masarinae). Z. Morphol. Ökol. Tiere **48,** 424—438 (1959).

Schuberg, A.: Über den Bau und die Funktion der Haftapparate des Laubfrosches. Arbeiten aus dem Zoologisch-Zootomischen Institut in Würzburg, Bd. 10, 57—118 (1895).

Schultze, E.: Die Angleichung der Verkehrsmittel an die Formen der Natur. Deutsche Technik, H. 4. Berlin-Leipzig 1937.

Schulz, E.: Über die Tardigraden. Zool. Anz. **171,** 3—12 (1963).

Schulze, F. E.: Epithel- und Drüsenzellen. Arch. mikr. Anat. **3,** 137—197 (1867).

Schumacher, S.: Zur Mechanik und Verwendungsart des Schnepfenschnabels. Z. Morphol. Ökol. **15,** 90—108 (1929).

Schwab, D. W., Shore, R. E.: Fine structure and composition of a siliceous sponge spicule. Biol. Bull. **140,** 125—136 (1971).

Schweinfurth, G.: Im Herzen von Afrika, Reisen und Entdeckungen etc. I. 1874.

Schwerdtfeger, F.: Autökologie. Die Beziehungen zwischen Tier und Umwelt. Hamburg-Berlin 1963.

Seilacher, A.: Arbeitskonzept zur Konstruktions-Morphologie. Lethaia **3,** 393—396 (1970).

Seling, Hrsg.: Keyser's Kunst- und Antiquitätenbuch II. München-Heidelberg: Keyser 1959.

Seydel, E.: Untersuchungen über den Byssusapparat der Lamellibranchiata. Zool. Jahrb. Anat. **27,** 465—582 (1909).

Siebold, Ph. Fr.: Fauna japonica. Leiden 1847—1849.

Simmermacher, G.: Haftapparate bei Wirbeltieren. Zool. Garten **25,** 289—301 (1884).

Simmermacher, G.: Haftapparate der Insekten. Z. Wiss. Zool. **40,** 481—556 (1884).

Simmermacher, G.: Untersuchungen über Haftapparate an Tarsalgliedern von Insekten. Z. Wiss. Zool. **40,** 481—456 (1884).

Simmermacher, G.: Antwort an Herrn Dr. Dewitz in Berlin. Zool. Anz. **7,** 513—517 (1884).

Simmermacher, G.: Untersuchungen über Haftapparate an Tarsalgliedern von Insecten. Zool. Anz. **7,** 225—228 (1884).

Sládeček, V.: Über die Exkretophoren bei den festsitzenden Rädertieren der Ordnung Collothecacea. Zool. Anz. **169,** 488—491 (1962).

Snodgrass, R. E.: The head and mouth parts of the cicada. Proc. Ent. Soc. Wash. **29,** 1—16 (1927).

Snodgrass, R. E.: The feeding apparatus of biting and sucking insects affecting man and animals. Smith. Misc. Coll. **104,** 1—113 (1944).

Sokol, O. M.: Studien an pipiden Fröschen. I. Die Kaulquappen von Hymenochirus curtipes Noble. Zool. Anz. **162,** 154—160 (1959).

Spielman, A.: The mechanics of copulation in Aëdes aegypti. Biol. Bull. **127,** 324—344 (1964).

St. Quentin, D.: Der Fangapparat der Odonaten. Österr. Zool. Z. **4,** 375—390 (1954).

Stammer, H. J.: „Trends" in der Phylogenie der Tiere; Ektogenese und Autogenese. Zool. Anz. **162,** 187—208 (1959).

Stanek, V. J.: Das große Bilderlexikon der Insekten. Prag: Artia 1968.

Steinbacher, G.: Die Zungenborsten der Loris. Zool. Anz. **146,** 57—65 (1951).

Steiner, G.: Untersuchungen über den allgemeinen Bauplan des Nematodenkörpers. Zool. Jahrb. Anat. **43,** 1—96 (1922).

Steinmann, P.: Die Tierwelt der Gebirgsbäche. Annales de Biologie lacustre, Tome II. Brüssel 1907.

Sterba, G.: Zur Morphologie und Biologie der Gattung Diplozoon. Zool. Anz. **158,** 181—196 (1957).

Storch, V.: Neue Polychaeten aus der Sandfauna des Roten Meeres. Zool. Anz. **178,** 102—110 (1967).

Storms, R.: The adhesive disk of Echineis. Ann. Mag. Nat. Hist. **2,** 65—76 (1888).

Strasburger, E.: Lehrbuch der Botanik. 30. Aufl. Stuttgart: Fischer 1971.

Strass, G.: Modell und Erkenntnis. Zur erkenntnis-theoretischen Bedeutung der Modellmethode in der Biologie. Jena: G. Fischer 1963.

Strasser, H.: Zur Kenntnis der funktionellen Anpassung. Stuttgart 1883.

Strasser, H.: Über den Flug der Vögel. Ein Beitrag zur Erkenntnis der mechanischen und biologischen Probleme der activen Locomotion. Jenaer Z. Naturwiss. **19,** 174—327 und 329—429 (1885).

Strenger, A.: Zur Kopfmorphologie der Ephemeridenlarven. 1. Teil: Ecdyonurus und Rhithrogena. Österr. Zool. Z. **4,** 191—228 (1954).

Strouhal, H.: Die Porcellionidae-Gattung Leptotrichus in der Türkei (Isop. terr.). Zool. Anz. **165,** 90—115 (1960).

Studer-Thiersch, A.: Die Flamingos. In Grzimeks Tierleben, Bd. VII, 1968.

Stümpke, H.: Bau und Leben der Rhinogradentia. Mit einem Nachwort von Gerolf Steiner. 12. bis 15. Tsd., Stuttgart 1972.

Sundermeier, W.: Der Hautpanzer des Kopfes und des Thorax von Myrmeleon europaeus und seine Metamorphose. Zool. Jahrb. Anat. **66,** 291—348 (1940).

Tait, N. N.: The anatomy of the Sawfly Perga Affinis Affinis Kirby (Hymenoptera: Symphyta). Australian. J. Zool. **10,** 652—683 (1962).

Tandler, J.: Beiträge zur Anatomie der Geckopfote. Z. Wiss. Zool. **75,** 308—326 (1903).

Thon, Ph. C. K.: Copulationsorgane der Hydrachniden-Gattung Arrhenurus. Verhandl. Deutsch. Zool. Gesellsch. 10. Vers. (1900).

Törne, O.: Die Saugnäpfe der männlichen Dytisciden. Zool. Jahrb. Anat. **29,** 415—448 (1910).

Tonnoir, A. L.: Descriptions of remarkable indian Psychodidae and their early stages, with a theory of the evolution of the ventral suckers of dipterous larvae. Rec. Ind. Mus. Calcutta **35,** 53—75

Tornier, G.: Ein Eidechsenschwanz mit Saugscheibe. Biol. Zentr. **19,** 549—552 (1899).

Traité de Zoologie: s. Grassé.

Tratz, E. P.: Vogelschnabel und Technik. Kosmos **19,** 200—203 (1922).

Tretzel, E.: Biologie, Ökologie und Brutpflege von Coelotes terrestris (Wider) (Araneae, Agelenidae). Teil I. Biologie und Ökologie. Z. Morphol. Ökol. Tiere **49,** 658—745 (1961).

Ulrich, H.: Der Fang- und Greifapparat von Mantispa – ein Vergleich mit Mantis. Natur und Museum **95,** 499—508 (1965).

Verhandelingen over de natuurlyke geschiedenis der Nederlandsche overzeesche Bezittingen. Zoologie. Leiden 1839—1844.

Virchow, H.: Die Mechanik der Schwanzwirbelsäule von Papio olivaceus. S.-B. Ges. Naturfreunde Berlin 181—86 (1914).

Wagner, F. v.: Zur Kenntnis des Baues der sog. Haftpapillen von Microstoma lineare Oerst. Zool. Anz. **14,** 327—331 (1891).

Weber, H.: Beiträge zur Bewegungsphysiologie der Hippocampus-Arten. Z. vergl. Physiol. **5,** 1—36 (1927).

Weber, H.: Zur vergleichenden Physiologie der Saugorgane der Hemipteren, mit besonderer Berücksichtigung der Pflanzenläuse. Z. vergl. Physiol. **8,** 145—186 (1929).

Weber, H.: Lebensweise und Umweltbeziehungen von Trialeurodes vaporariorum (Westwood) (Homoptera-Aleurodina). Erster Beitrag zu einer Monographie dieser Art. Z. Morphol. Ökol. Tiere **23,** 575—753 (1931).

Weber, H.: Vergleichend funktionsanatomische Untersuchungen an atypischen Beißmandibeln von Insekten mit besonderer Berücksichtigung der Phyllotoma-Larve (Hymenopt., Tenthred.). Biol. Zentr. **59,** 541—566 (1939).

Weber, H.: Grundriss der Insektenkunde. 2. überarbeitete Aufl. Jena 1949. 3. überarbeitete Aufl. Stuttgart 1954.

Weber, H.: Konstruktionsmorphologie. Zool. Jahrb. Abt. Allg. Zool. **68,** 1—112 (1958).

Weinreich, E.: Über den Klebfangapparat der Imagines von Stenus Latr. (Coleopt., Staphylinidae) mit einem Beitrag zur Kenntnis der Jugendstadien dieser Gattung. Z. Morphol. Ökol. Tiere **62,** 162—210 (1968).

Weitlander, F.: Eine Untersuchung über den Haftfuß des Gecko. Verhandl. Zool.-Bot. Ges. Wien **52,** 328—332 (1902).

Wendler, G.: Über die Fortbewegung der Larven von Cantharis fusca. Z. vergl. Physiol. **48,** 283—294 (1964).

Wendler, G.: Über einige Modelle in der Biologie. Stud. gen. Berlin **18,** 284—290 (1965).

Wesché, W.: Genitalia of both the sexes in Diptera. Trans. Linn. Soc. London. 2. Ser. **9** (1906).

Wesenberg-Lund, C.: Fortpflanzungsverhältnisse der Süßwasserinsekten Fortschr. Nat. Wiss. Forsch. **8** (1913).

Wesenberg-Lund, C.: Odonaten-Studien. Int. nat. Revue Hydrob. Plankt. **6** (1913).

Wesenberg-Lund, C.: Biologie der Süsswassertiere. Wirbellose Tiere. Wien: Springer 1939.

Wesenberg-Lund, C.: Biologie der Süßwasserinsekten. Wien Springer 1943.

West Tuffen: The foot of the fly; its structure and action: elucidated by comparison with the feet of other Insects etc. Part I. In: Transactions of the L. Society of London. Vol. XXIII. London p. 393—421 Tab. XLI-XLIII.

Wettstein, A.: Über die Fischfauna des tertiären Glarnerschiefers. Untersuchungen aus der geologischen Sammlung in Zürich. Abhandl. Schweiz. Paläontol. Ges. **13** (1887).

Wetzel, A.: Beitrag zur Kenntnis der Trichodina pediculus Elerbg, (Infusoria peritricha).

Wetzel, A.: Studien über die Biologie der Caprelliden. I. Bewegung, Nahrungserwerb, Aufenthaltsort. Z. Wiss. Zool. **141,** 347—398 (1932).

Wheeler, W. M.: Arthropods: Their structure, development and behaviour. New York 1960.

Wheeler, R. E., Jones, J. C.: The mechanics of copulation in Aedes aegypti mosquitoes. Anat. Rec. **138,** 388 (1960).

Wickler, W.: Weitere Untersuchungen über Haftfäden an Teleosteer-Eiern, speziell an Cyprinodon variegatus Lacépède 1803. Zool. Anz. **163,** 90—107 (1959).

Wiehle, H.: Verschiedene Fanggewebe einheimischer Spinnen. Leipzig 1969.

Wigley, R. L., Shave, P.: Caprella grahami, a new species of Caprellid (Crustacea: Amphipoda) commensal with Starfishes. Biol. Bull. **130,** 289—296 (1966).

Williamson, E. B.: Copulation of Odonata. Entomol. News. vol. 17.

Wittich, v.: Der Mechanismus der Haftzehen von Hyla arborea. Arch. Anat. Physiol. (1854).

Yanagita, T. M.: VII. Extrusion of resting cnidae—its nature and its possible bearing on the normal nettling response. J. Exp. Biol., Edinburgh **36** (3), 418—494 (1959c).

Zander, E.: Männliche Geschlechtsanhänge der Hymenopteren. Z. Wiss. Zool. **67** (1900).

Zander, E.: Männliche Geschlechtsanhänge der Trichopteren. Z. Wiss. Zool. **70** (1901).

Zander, E.: Morphologie der männlichen Geschlechtsanhänge der Lepodopteren. Z. Wiss. Zool. **74** (1903).

Zander, E.: Der männliche Genitalapparat der Butaliden. Z. Wiss. Zool. **79** (1905).

Zick, K.: Urceolaria Korschelti n. sp., eine neue marine Urceolarine, nebst einem Überblick über die Urceolarinen. Z. Wiss. Zool. **132** (Korschelt-Festband), 355—403 (1928).

Ziswiler, V.: Zur Kenntnis des Samenöffnens und der Struktur des hörnernen Gaumens bei körnerfressenden Oscines. J. Ornithol. **106,** 1—48 (1965).

Subject and Author Index

Page numbers in italics refer to figures

GPSR Compliance

The European Union's (EU) General Product Safety Regulation (GPSR) is a set of rules that requires consumer products to be safe and our obligations to ensure this.

If you have any concerns about our products, you can contact us on ProductSafety@springernature.com

In case Publisher is established outside the EU, the EU authorized representative is:

Springer Nature Customer Service Center GmbH
Europaplatz 3
69115 Heidelberg, Germany

Batch number: 10620881

Printed by Printforce, the Netherlands